Yield-line formulae for slabs

K. W. Johansen

Yield-line formulae for slabs

K. W. Johansen

Professor, DrTechn,
Technical University of Denmark

Cement and Concrete Association

The cost of translating this standard work by Professor Johansen has been met by a group of his friends among Danish consulting engineers

Translated by Pauline M. Katborg
from the third (1968) edition of the Danish book *Pladeformler*
published by Polyteknisk Forlag, Copenhagen

Price : £3·00

SBN 7210 0819 4

12.044 First published 1972

Designed and published by the
Cement and Concrete Association,
52 Grosvenor Gardens, London, SW1W 0AQ

Printed Uneoprint by
Unwin Brothers Limited, The Gresham Press,
Old Woking, Surrey.
A member of the Staples Printing Group (UO1940)

Contents

Foreword

On 26 August 1971, Professor K. W. Johansen, Dr techn, celebrates his seventieth anniversary, and this means that he will be retiring after forty years of inspiring work at the Structural Research Laboratory of the Technical University of Denmark.

K. W. Johansen has been a member of the board of the Danish Society for Structural Science and Engineering for many years, with two long periods as Chairman; in 1970, he was made an honorary member of the Society. Since 1949, he has been a member of the Danish Academy of Technical Science. In 1955, he received the Augustinus Award, a scientific grant conferred by a foundation, the administrative board of which includes the vice-chancellors of three Danish universities, and in 1971, the Institution of Structural Engineers, London, honoured him with its Gold Medal.

Internationally, Professor Johansen is known for his work in the International Association for Bridge and Structural Engineering, and for several years he has been a member of the Executive Council of the International Association for Shell Structures. Although, in international circles, our retiring professor is usually referred to as Professor Johansen, among Danish colleagues he is best known as K. W.

K. W's first paper on the ultimate bearing capacity of reinforced concrete slabs was "Beregning af krydsarmerede jernbetonpladers brudmoment" (Calculation of the yield moment of cross-reinforced concrete slabs), Bygningsstatiske Meddelelser, Vol. 3, 1931, pp. 1-18. The German version of this paper, "Bruchmomente der kreuzweise bewehrten Platten", was published in the International Association for Bridge and Structural Engineering's "Publications", Vol. 1, 1932, pp. 277-296. In the same year, Johansen published another paper on slabs, "Nogle pladeformler" (Some formulae for slabs), in Bygningsstatiske Meddelelser, Vol. 4, 1932, pp. 77-84.

K. W. Johansen's yield-line theory for slabs was adopted very early by Danish engineers. This was possible because of Denmark's very liberal traditions as regards codes of practice. As long ago as 1908, the Danish codes for reinforced concrete structures specified a beam analysis that can be characterized as a modified yield-hinge theory, and in 1921, the yield-line theory for slabs proposed by the Danish engineer Aage Ingerslev came into use.

In 1943, K. W. Johansen published his thesis, "Brudlinieteorier" (Yield-line theories), which resulted in increasing use of his method. Unfortunately, only a few engineers outside Scandinavia were able to read Johansen's thesis because it was in Danish; some engineers even went to the trouble of learning Danish in order to get the full benefit of this important book. Fortunately, in 1962, it was finally translated into English and Portuguese ("Yield-line theory", London, Cement and Concrete Association, 1962; "Linhas de ruptura; teoria e prática", Rio de Janeiro, Ao Livro Tecnico S. A., 1962).

In the meantime, Johansen published further papers in which he developed a large number of formulae for slabs: "Pladeformler" (Formulae for slabs). These were first published in 1946. Because they are easy to apply in practice, these formulae have been widely used in Denmark ever since. A certain amount of practice is necessary in using the yield-line theory because the type of yield-line pattern to be considered has to be estimated in each case. The present book, which is a translation of Johansen's Danish book, "Pladeformler", should eliminate this obstacle for all the cases it covers. However, a study of the original thesis, "Yield-line theory", is recommended for those wishing to derive full benefit from the present book.

One great advantage of the yield-line theory is that it covers many different shapes of slab and complicated cases of loading. It is a difficult problem to analyse the elastic stress distribution in such cases, whereas a solution can usually be reached quite easily by means of the yield-line theory.

Johansen's papers cover the entire field of the theory of structures, elasticity, plasticity, stability, dynamic problems, plates, shells, folded plates, steel, reinforced concrete, prestressed concrete, timber, etc. This is amply illustrated by the bibliography in this book, which covers sixty of his papers.

Most engineers know Johansen for his work on yield-line theory. He sometimes finds this a little depressing, remarking that people seem to think he knows little or nothing else, and that he is thoroughly fed up with his theory. We were, therefore, somewhat apprehensive about suggesting that this English edition of his "Formulae for slabs" should be published, but fortunately our idea met with his approval.

As most of Johansen's other works have only been published in Danish, we would like to mention a few of the most important papers here.

As early as 1928, he wrote "Om virkningen af bøjler og skraajern i jernbetonbjælker" (On the effect of stirrups and bent-up bars in reinforced concrete beams), in which he presented an original discussion of the action of reinforcement in concrete beams, which departed completely from the conventional views of the time. The same questions were touched upon in his paper, "Beregning af jernbetonbjælker" (Design of reinforced concrete beams), Bygningsstatiske Meddelelser, Vol. 16, 1945, pp. 35-64, and in his contribution, "Critical remarks on the effect of bent-up bars and stirrups in reinforced concrete beams", made to International Association for Bridge and Structural Engineering, Fifth Congress, Lisbon-Oporto, 1956, Final Report 1957, pp. 507-512. Some of the ideas presented in these papers differ very little from those adopted at a much later date in codes and recommendations.

In 1937, Johansen wrote a paper, "Bøjningsfri spændingstilstande i skaller" (Membrane stresses in shells), and in 1944, "Skalkonstruktion paa Radiohuset. Beregning og forsøg" (Shell at the Radio Centre. Design and tests). In the latter he presented a theory of rupture for cylindrical shells. The method was used for the design of a large roof structure in Copenhagen, and his ideas were later adopted and further developed by H. Lundgren in his thesis, "Cylindrical shells", Copenhagen, The Danish Technical Press, 1949.

Johansen's paper, "Pladetrapper" (Staircase slabs) from 1940 provides a good example of his simple approach to practical problems—in this case, folded-plate structures.

Although Johansen is a very fine mathematician, he disapproves of the use of complicated mathematical solutions in cases in which a sound engineering approach can solve the problem. He constantly emphasizes the importance of the real physical behaviour of the materials and of simple and logical analysis. He is known to be very critical, but his contributions to discussions have always held the attention of his audience. We cannot resist the temptation here to quote his comment on safety from "Forspændt beton" (Prestressed concrete), Copenhagen, 1951:

"It should be noted that the antiquated concept 'permissible stresses' has been completely abandoned in the foregoing considerations. We are left with the concept of safety, which is generally settled more or less conventionally in the form of a factor of safety, the magnitude of which is usually agreed upon by a suitable commission of nice, elderly engineers. The factors of safety are preferably selected from the positive integers higher than unity, although such values as 1·5 and 2·5 are also encountered."

Troels Brøndum-Nielsen
on behalf of the
Structural Research Laboratory of the
Technical University of Denmark

List of K. W. Johansen's scientific publications

The papers are arranged in chronological order. Papers published within the same year are arranged in alphabetical order.

Most of the papers are in Danish. The language of each particular paper is that in which the title is first given.

Nomogram til beregning af T-bjælker med tynd plade uden hensyntagen til kroppens trykspændinger. (Nomogram for the analysis of T beams with thin plate without regard to the compressive stresses in the web.) *Ingeniøren*. Vol. 35, No. 9. February 1926. pp. 101-103.

Symmetriske jærnbetontværsnit. (Reinforced concrete beams with symmetrical sections.) *Ingeniøren*. Vol. 36, No. 45. November 1927. pp. 542-546.

Om virkningen of bøjler og skråjern i jernbetonbjælker. (On the effect of stirrups and bent-up bars in reinforced concrete beams.) Copenhagen, Danmarks Naturvidenskabelige Samfund, 1928. pp. 52. A No. 17.

Bjælkesøjler. (Beam-columns.) *Bygningsstatiske Meddelelser*. Vol. 2, No. 5. October 1930. pp. 85-94.

Styrkeforholdene i støbeskel i beton. (The strength of concrete in construction joints.) *Bygningsstatiske Meddelelser*. Vol. 2, No. 4. August 1930. pp. 67-68.

Søjler med variabel kraft og stivhed. (Columns with variable axial force and rigidity.) *Bygningsstatiske Meddelelser*. Vol. 2, No. 2. March 1930. pp. 23-38.

Beregning af krydsarmerede jernbetonpladers brudmoment. (Calculation of the yield moment of cross-reinforced concrete slabs.) *Bygningsstatiske Meddelelser*. Vol. 3, No. 1. February 1931. pp. 1-18.

Bruchmomente der kreuzweise bewehrten Platten. (Yield moments in cross-reinforced slabs.) Zürich, International Association for Bridge and Structural Engineering, 1932. *Publications*. Vol. 1. pp. 277-296.

Nogle pladeformler. (Some formulae for slabs.) *Bygningsstatiske Meddelelser*. Vol. 4, No. 4. December 1932. pp. 77-84.

Bøjningsfri spændingstilstande i skaller. (Membrane stresses in shells.) *Bygningsstatiske Meddelelser*. Vol. 9, No. 2. 1937. pp. 61-84.

Pladetrapper. (Staircase slabs.) *Bygningsstatiske Meddelelser*. Vol. 11, No. 2. 1940. pp. 13-21.

Ritz's tilnærmelsesmetode. (The approximate method of Ritz.) *Bygningsstatiske Meddelelser*. Vol. 11, No. 2. 1940. pp. 25-26.

Specielle trækonstruktioner. (Special timber structures.) (Co-author: J.-E. Ekström). In Willy Hansen (Ed.): *Tømrer- og bygningssnedkerarbejde*. Copenhagen, Emil Wienes Bogforlag, 1940. pp. 232-272.

Strengbeton. (Pre-tensioned concrete.) *Bygningsstatiske Meddelelser*. Vol. 11, No. 2. 1940. pp. 22-24.

Udmattelsesfænomener. (Fatigue problems.) *Bygningsstatiske Meddelelser*. Vol. 11, No. 2. 1940. pp. 31-32.

Vridning af trappetrin. (Torsion of stairs.) *Bygningsstatiske Meddelelser*. Vol. 11, No. 2. 1940. pp. 27-30.

Brudlinieteorier. (Yield-line theories.) Thesis. Copenhagen, Jul. Gjellerups Forlag, 1943. pp. 191. Reprinted 1952, 1958 and 1961.

Superpositionsloven og Hooke's lov. (The law of superposition and Hooke's law.) *Bygningsstatiske Meddelelser*. Vol. 14, No. 2. 1943. pp. 53-58.

Skalkonstruktion paa Radiohuset. Beregning og forsøg. (Shell at the Radio Centre. Design and tests.) *Bygningsstatiske Meddelelser*. Vol. 15, No. 1. 1944. pp. 1-26.

Beregning af jernbetonbjælker. (Design of reinforced concrete beams.) *Bygningsstatiske Meddelelser*. Vol. 16, No. 2. 1945. pp. 35-64.

Laboratoriet for Bygningsstatik: Indmurede trapper. (Built-in stairs.) *Bygningsstatiske Meddelelser*. Vol. 16, No. 1. 1945. pp. 27-34.

Forsøg med hulstensdæk. (Tests with hollow-pot floors.) *Bygningsstatiske Meddelelser*. Vol. 17, No. 1. 1946. pp. 18-44.

Pladeformler. (Formulae for slabs.) Copenhagen, Polyteknisk Forening, 1946. pp. 186. Second edition, Copenhagen, Polyteknisk Forening, 1949. pp. 186. Third edition, Copenhagen, Polyteknisk Forlag, 1968. pp. 240.

Vridningsforsøg med trappetrin. (Torsion tests with stairs.) *Bygningsstatiske Meddelelser*. Vol. 17, No. 1. 1946. pp. 11-17.

Plader på jord. (Plates on soil.) *Dansk Vejtidsskrift*. Vol. 26, No. 8. August 1949. pp. 146-157.

Theory of timber connections. Zürich, International Association for Bridge and Structural Engineering, 1949. *Publications*. Vol. 7. pp. 249-262.

Critical notes on the calculation and design of cylindrical shells. *International Association for Bridge and Structural Engineering. Third Congress, 1948, Liège. Final Report, 1950.* pp. 601-606.

Forelæsninger over elasticitets- og styrkelære, 1-5. (Lectures on the theory of elasticity and strength of materials.) Copenhagen, Polyteknisk Forening, 1950-53. pp. 125. Second edition, Copenhagen, Polyteknisk Forening, 1958-59. pp. 161. Third edition (abbreviated edition in one volume), Copenhagen, Akademisk Forlag, 1967. pp. 96.

Hjultryk på indspændt plade med fri forstærket rand. (Concentrated loads on built-in slabs with a free, strengthened edge.) *Bygningsstatiske Meddelelser*. Vol. 21, No. 5. 1950. pp. 145-160.

Om differensligninger. (On difference equations.) Copenhagen, Structural Research Laboratory, Technical University of Denmark, 1950. Bulletin No. 1. pp. 64-70.

On integration of the differential equation for thin shells without bending. *International Association for Bridge and Structural Engineering, Third Congress, 1948, Liège. Final Report, 1950.* pp. 597-600.

The ultimate strength of reinforced concrete slabs. *International Congress for Bridge and Structural Engineering. Third Congress, 1948, Liège. Final Report, 1950.* pp. 565-570.

Forspændt beton. (Prestressed concrete.) In: Dansk Ingeniørforening (Ed.): *Forspændt beton.* Copenhagen, Teknisk Forlag, 1951. pp. 37-44.

Ekscentrisk påvirkede nittegrupper. (Eccentrically loaded groups of rivets.) *Bygningsstatiske Meddelelser*. Vol. 23, No. 2. 1952. pp. 17-22.

Bjælker og buer. (Beams and arches.) Copenhagen, Structural Research Laboratory, Technical University of Denmark, 1954. Bulletin No. 3. pp. 33-52.

?lasticitetsteorien for plankrumme bjælker belastede vinkelret på deres plan. The theory of plasticity for plane-curved beams loaded at right-angles to :heir plane.) Copenhagen, Structural Research Laboratory, Technical University)f Denmark, 1954. Bulletin No. 3. pp. 33-52.

Skæv bøjning. (Unsymmetrical bending.) Copenhagen, Structural Research Laboratory, Technical University of Denmark, 1954. Bulletin No. 3. pp. 21-32.

Forelæsninger over svingningsteori. Faseplanmetoden. (Lectures on vibration theory. The phase plan method.) Copenhagen, Polyteknisk Forening, 1955. pp. 51.

Critical remarks on the effect of bent-up bars and stirrups in reinforced concrete beams. *International Association for Bridge and Structural Engineering. Fifth Congress, Lisbon-Oporto, 1956. Final Report, 1957.* pp. 507-512.

Matricer. Teori. Praksis. (Matrices. Theory. Practice.) Copenhagen, Polyteknisk Forening, 1957. pp. 128.

Matricer. Formelsamling. (Matrices. Formulae.) Copenhagen, Polyteknisk Forening, 1957. pp. 19.

Et par åbenhjertige ord i anledning af en 100-årsdag. (A few plain words on the occasion of a centenary.) *Ingeniøren.* Vol. 66, No. 29. November 1957. pp. 805-806.

Brudbetingelser for sten og beton. (Criteria of strength for stone and concrete.) *Bygningsstatiske Meddelelser.* Vol. 29, No. 2. November 1958. pp. 25-44. Discussion, ibid. Vol. 30, No. 2. September 1959. pp. 88-89.

Shell roof construction. Discussion. *Proceedings of the Second Symposium on Concrete Shell Roof Construction, 1-3 July, 1957.* Oslo, Teknisk Ukeblad, 1958. pp. 101, 270-272, 336-337.

Strength of concrete under combined tensile and compressive stresses. Discussion. *Journal of the American Concrete Institute. Proceedings* Vol. 54, No. 6, Part 2. December 1958. pp. 1303-1305.

Nogle formler for laveste svingningstal ved bøjningssvingninger. (Some formulae for the lowest frequency of transverse vibration of beams.) Copenhagen, Structural Research Laboratory. Technical University of Denmark, 1959. Bulletin No. 10. pp. 139-151.

Strength of concrete under combined stresses. Discussion. *Journal of the American Concrete Institute. Proceedings* Vol. 55, No. 9. March 1959. pp. 1043-1045.

An iterative method of calculation for axial symmetric plates with bending—as well as membrane stresses. (Co-author: C. Dyrbye.) *Bygningsstatiske Meddelelser.* Vol. 31, No. 3. December 1960. pp. 77-92.

Svingningsteori I-IV. (Theory of vibration.) Copenhagen, Polyteknisk Forening, 1961-1964. pp. 230. Second edition, Copenhagen, Akademisk Forlag, 1964-65. pp. 267. Third edition (in one volume), Copenhagen, Akademisk Forlag, 1969. pp. 246.

Linhas de ruptura. Teoria e prática. (Translation into Portuguese by J. Mason of parts of: *Brudlinieteorier,* 1943, and *Pladeformler,* 1946). Rio de Janeiro, Ao Livro Tecnico S.A., 1962. pp. 380.

Yield-line theory. London, Cement and Concrete Association, 1962. pp. 181.

The yield-line theory. Notes for lectures, 16 March 1962. Institution of Structural Engineers, Lancashire and Cheshire Branch. pp. 16.

Forskningen. (Research.) *Ingeniøren.* Vol. 72, No. 11. June 1963. pp. 380-382.

Limit analysis of reinforced concrete shells. In: Olszak, W. and A. Sawczuk (Ed.) *Non-classical shell problems: Proceedings, IASS symposium, Warsaw, September 1963.* Amsterdam, North Holland Publishing Company and Warsaw, PWN-Polish Scientific Publishers, 1964. pp. 937-940.

Plastisite teorisinin uygulanmasi üserine düsünceler. Teknik brosür. (Translation into Turkish of: Beregning og dimensionering, 1969; Theory of timber connections, 1949; Ekscentrisk påvirkede nittegrupper, 1952; Pladetrapper, 1940. Translated by G. Özmen.) Istanbul, Insaat Mühendisleri Odasi Yayinlari, 1967. pp. 44.

Anvendelser af energiprincipperne. Tilnærmelsesmetoder. (Application of the energy theorems. Approximations.) Copenhagen, Laboratoriet for Bærende Konstruktioner, Danmarks Tekniske Højskole, 1969. pp. 19. Forelæsningsnotater nr. 7.

Beregning og dimensionering. (Analysis and design.) Copenhagen, Laboratoriet for Bærende Konstruktioner, Danmarks Tekniske Højskole, 1969. pp. 13. Forelæsningsnotater nr. 14.

Energi- og arbejdsprincipper. (Energy and work theorems.) Copenhagen, Laboratoriet for Bærende Konstruktioner, Danmarks Tekniske Højskole, 1969. pp. 19. Forelæsningsnotater nr. 3.

Foldning af plader. (Buckling of plates.) Copenhagen, Laboratoriet for Bærende Konstruktioner, Danmarks Tekniske Højskole, 1969. pp. 20. Forelæsningsnotater nr. 8.

Symmetri. (Symmetry.) Copenhagen, Laboratoriet for Bærende Konstruktioner, Danmarks Tekniske Højskole, 1969. pp. 11. Forelæsningsnotater nr. 10.

Stabilitetsproblemer. (Stability problems.) Copenhagen, Afdelingen for Bærende Konstruktioner, Danmarks Tekniske Højskole, 1971. pp. 26. Forelæsningsnotater nr. F28.

Introduction

In this book a first attempt is made to establish simple formulae for the calculation of various types of slab that are of interest in practice. The book includes a large number of examples of the application of the yield-line theory for slabs, as presented in my book "Yield-line theory", which will be referred to throughout this book by the abbreviation "YLT".

The same notation and sign convention are used as in YLT. As regards the term "restraint", the following remarks should be noted. It is necessary to differentiate between geometric and static restraint. With geometric restraint, movements of the slab are completely prevented at the restraint and the restraining moment is statically indeterminate. With static restraint, the restraining moment is given and is equal to the ultimate moment in the section of restraint. The movement of the section of restraint is geometrically indeterminate. The older term, partial restraint, corresponds more closely to static restraint, although it does not cover this completely, as partial restraint can only give restraining moments that are smaller than those corresponding to geometric restraint, whereas static restraint may well result in greater restraining moments. In the following, restraint should be taken to mean static restraint, unless otherwise stated.

The true value of the ultimate moment m corresponds to the failure load. However, in practical calculations, only a fraction of this—the working load—is used, and the corresponding value of m should then really have another name, for example, yield-line moment, which is the same fraction of the true ultimate moment. However, this distinction is not made in the following, the term ultimate moment or, simply, the moment m being used instead of yield-line moment, so the possibility of mistakes is precluded.

Decimal classification is used throughout, both for chapters and sub-sections and for the corresponding formulae and figures.

0 Slabs spanning in one direction (beams)

0.1 Uniformly distributed load

A continuous slab (beam) with a uniformly distributed total load (live load plus dead load) q per unit length usually yields at the supports and also in the span. The positive ultimate moment (see Fig. 0.11) being m and the negative ultimate moment being $-m'$, we have at m that the shear force $Q = 0$, as the moment is maximum.

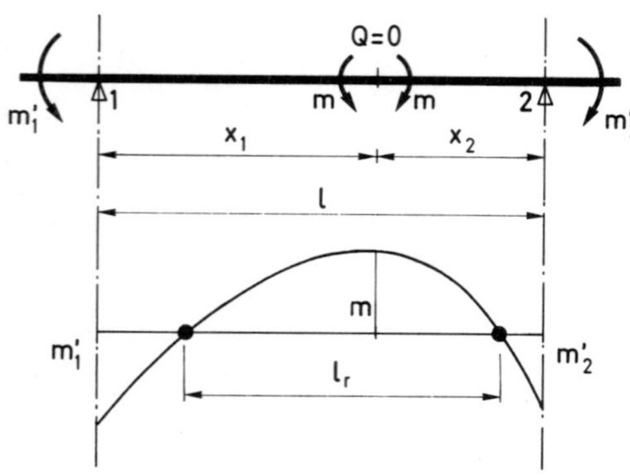

Fig. 0.11

The moment equations about 1 and 2 respectively yield

$$m + m'_1 = \tfrac{1}{2}qx_1^2$$

and

$$m + m'_2 = \tfrac{1}{2}qx_2^2$$

Defining the *reduced span* l_r by means of $m = \tfrac{1}{8}ql_r^2$ and the *degrees of restraint* by $i_1 = m'_1/m$; $i_2 = m'_2/m$, we get

$$l_r^2(1 + i_1) = 4x_1^2, l_r^2(1 + i_2) = 4x_2^2$$
$$2x_1 + 2x_2 = 2l = l_r(\sqrt{1 + i_1} + \sqrt{1 + i_2}) \quad (0.11)$$

It will be seen that l_r is the distance between the points of zero moment. The reactions from the span under consideration will be

$$R_1 = qx_1 = \tfrac{1}{2}ql_r\sqrt{1 + i_1}$$
$$R_2 = qx_2 = \tfrac{1}{2}ql_r\sqrt{1 + i_2} \quad (0.12)$$

In practice, m'_1 and m'_2 will usually be m or 0, so l_r/l becomes $2/(1 + \sqrt{2}) = 0{\cdot}83$ and $1/\sqrt{2} = 0{\cdot}71$ in the outer and the intermediate span, respectively. By dividing the slab so that the outer span is 0·85 of the intermediate spans, the same l_r is obtained in all spans. If m_o is the simple moment $\tfrac{1}{8}ql^2$, we get $m/m_o = l_r^2/l^2$, which is 0·69 and 0·50 in the outer span and intermediate span, respectively.

By applying the common approximation

$$m = m_o - \tfrac{1}{2}(m'_1 + m'_2),$$

we get $\tfrac{2}{3}$ and $\tfrac{1}{2}$, respectively, for m/m_o, so this can perfectly well be used, i.e.

$$m = \frac{ql^2}{4(2 + i_1 + i_2)} \quad (0.13)$$

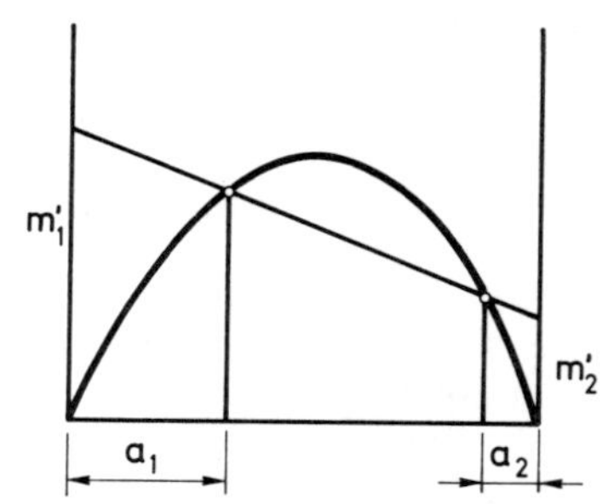

Fig. 0.12

By intersecting between the moment parabola $\tfrac{1}{2}gx(l - x)$ for dead load and the straight line $m'_1[(l - x)/l] + m'_2(x/l)$, we obtain the limits for the top reinforcement.

$$\frac{a_1}{l} = \frac{1}{2} + \frac{m'_1 - m'_2}{gl^2} - \sqrt{\frac{1}{4} - \frac{m'_1 + m'_2}{gl^2} + \left(\frac{m'_1 - m'_2}{gl^2}\right)^2}$$

$$\frac{a_2}{l} = \frac{1}{2} + \frac{m'_2 - m'_1}{gl^2} - \sqrt{\frac{1}{4} - \frac{m'_1 + m'_2}{gl^2} + \left(\frac{m'_1 - m'_2}{gl^2}\right)^2} \quad (0.14)$$

In practice, we usually have $i_1 = i_2 = i$ in intermediate spans and $i_1 = i$, $i_2 = 0$ in outer spans. Thus, from (0.13) and (0.14), we get:

For outer spans:

$$m = \frac{ql^2}{4(2 + i)}, \frac{a_1}{l} = \frac{i}{2(2 + i)}\frac{q}{g}, a_2 = 0 \quad (0.15)$$

For intermediate spans:

$$m = \frac{ql^2}{8(1 + i)}, \frac{a_1}{l} = \frac{a_2}{l} = \frac{1}{2}\left[1 - \sqrt{1 - \frac{i}{1 + i}\frac{q}{g}}\right] \quad (0.16)$$

It will be seen from this that in the intermediate

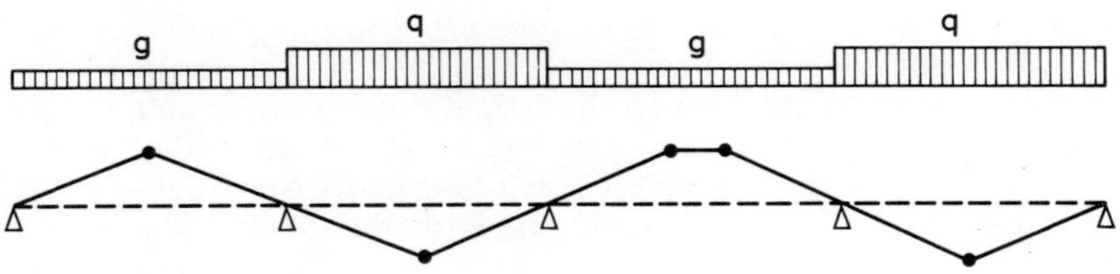

Fig. 0.13

spans the top reinforcement must be carried through the entire span when

$$\frac{q}{g} \geqslant 1 + \frac{1}{i} \qquad (0.17)$$

If the top reinforcement in the intermediate spans is only to be taken as far as the quarter-points, we put $a_1/l = 1/4$ in equation (0.16) and find that

$$i \leqslant \frac{3g}{4q - 3g} \qquad (0.18)$$

Then, in the outer spans,

$$\frac{a_1}{l} \leqslant \frac{3q}{2(8q - 3g)} \qquad (0.19)$$

If these rules are not observed, the slab will fail by cracking at the top, just at the ends of the top reinforcement, and it will not be possible for the yielding at the intermediate supports assumed in the calculation to take place (Fig. 0.13).

0.2 Point loads

When a point load, P, traverses the slab (beam), the apex of the triangular moment diagram describes the parabola $(P/l)\,x(l - x)$, which corresponds to the moment curve for the uniformly distributed load $2P/l$. In the most dangerous position, the load is, of course, located at the yield section, and the moment becomes

$$m = \frac{1}{8}\left(q + \frac{2P}{l}\right) l_r^2 \qquad (0.21)$$

The approximation (0.13) can, of course, also be used when q is increased by $2P/l$.

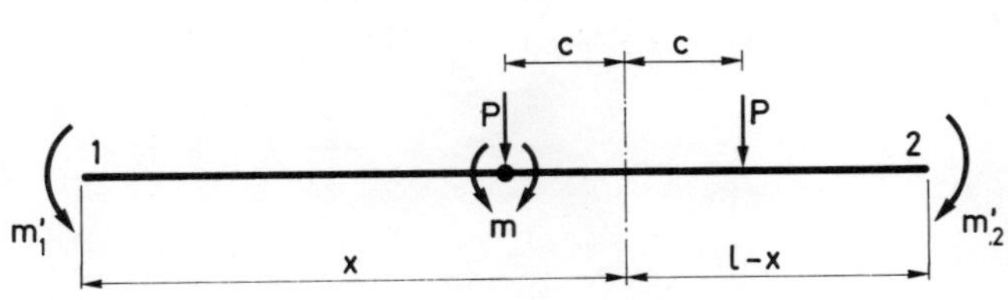

Fig. 0.21

According to the work equation, two forces of equal magnitude P, at a distance 2c from each other, give

$$(m + m_1')\frac{\delta}{x - c} + (m + m_2')\frac{\delta}{l - x + c} = P\delta + P\,\frac{l - x - c}{l - x + c}\,\delta$$

or

$$m = \frac{2P}{l}(l - x)(x - c) - m_1'\,\frac{l + c - x}{l} - m_2'\,\frac{x - c}{l}$$

assuming that failure occurs under the load nearest 1, where the deflexion is δ.

This shows that m can be determined by means of the free-moment diagram due to the uniformly distributed load $4P/l$ on a beam from $x = c$ to $x = l$ and a support-moment diagram that has the ordinates m_1' and $m_1'(c/l) + m_2'[(l - c)/l]$ at $x = c$ and $x = l$, i.e. the degrees of restraint $i_c = i_1$ and $i_l = i_1(c/l) + i_2[(l - c)/l]$. We thus obtain the reduced span

$$l_c = \frac{2(l - c)}{\sqrt{1 + i_1} + \sqrt{1 + i_2 - (i_2 - i_1)\frac{c}{l}}},\ (i_2 \geqslant i_1) \qquad (0.22)$$

and

$$m = \frac{1}{8}\,\frac{4P}{l}\,l_c^2 \qquad (0.23)$$

If the failure occurs under the other force, then $i_1 > i_2$, since these must be interchanged in (0.22) and the value of l_c thereby determined must be greater than the previous value.

For intermediate spans, where $i_1 = i_2 = i$, we therefore get $l_c = (l - c)/\sqrt{1 + i}$, and in outer spans, where $i_1 = 0, i_2 = i, l_c = 2(l - c)/[1 + \sqrt{1 + i - i(c/l)}]$.

If the approximation (0.13) is used, it will prove to be correct for intermediate spans, but for outer spans it will not prove to be as good in the case of point loads as in that of uniformly distributed loads.

As is known, a single wheel load will be more dangerous when c exceeds a certain value. This value can be found from (0.21) and (0.23):

$$\frac{1}{8}\,\frac{2P}{l}\,l_r^2 = \frac{1}{8}\,\frac{4P}{l}\,l_c^2,\ l_r = \sqrt{2}\,l_c$$

From this, by using (0.11) and (0.22), we find that for intermediate spans:

$$c = l\left(1 - \sqrt{\frac{1}{2}}\right) = 0{\cdot}293\ l$$

and, similarly, for outer spans:

$$1 + \sqrt{1 + i - i\frac{c}{l}} = \sqrt{2}(1 + \sqrt{1 + i}\,)\left(1 - \frac{c}{l}\right)$$

which gives the above value for $i = 0$ and $c/l = 0{\cdot}328$ for $i = 1$. The variation is of no practical importance so we can always reckon with $c = 0{\cdot}3\ l$.

The foregoing shows that the calculation of restrained slabs can be converted to the calculation of simply supported slabs by reducing the span to l_r. Analogously, the calculation of slabs subjected to point loads can be converted to the calculation of slabs subjected to a uniformly distributed load $2P/l$ for one load and $4P/l$ for two loads. The reduced span is l_c.

The following will show that corresponding procedures can be adopted for various types of two-way slab. In these cases, the line load $\bar{p}$ corresponds to the point loads.

1 Rectangular slabs

1.1 Supported on all four sides

1.11 Uniformly distributed load
Let p be the constant load per unit area. The sides are assumed to be restrained with the restraining moments $m_1' = i_1 m$, $m_2' = i_2 m$, $m_3' = i_3 m$ and $m_4' = i_4 m$. For $i = 0$ we get simple support.

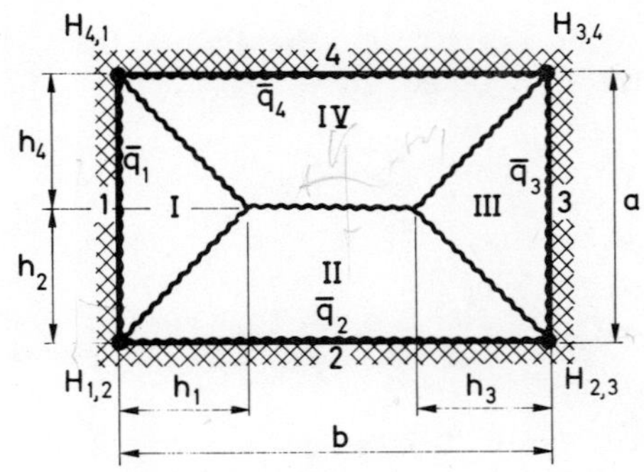

Fig. 1.111

The moment equations for I and III in the yield pattern shown in Fig. 1.111 give

$$\text{I} \quad m + m_1' = m(1 + i_1) = \tfrac{1}{6}\, p h_1^2$$

$$\text{III} \quad m + m_3' = m(1 + i_3) = \tfrac{1}{6}\, p h_3^2$$

For II and IV, we obtain from equation (36) in YLT:

$$\text{II} \quad m + m_2' = m(1 + i_2) = \frac{1}{6}\, p h_2^2 \left(3 - 2\,\frac{h_1 + h_3}{b}\right)$$

$$\text{IV} \quad m + m_4' = m(1 + i_4) = \frac{1}{6}\, p h_4^2 \left(3 - 2\,\frac{h_1 + h_3}{b}\right)$$

From these four equations and $h_2 + h_4 = a$, we can determine m, h_1, h_2, h_3 and h_4 the i's being assumed to be given.

From II and IV we find that

$$\frac{h_2}{\sqrt{1 + i_2}} = \frac{h_4}{\sqrt{1 + i_4}} = \frac{a}{\sqrt{1 + i_2} + \sqrt{1 + i_4}} = \frac{a_r}{2} \qquad (1.111)$$

where a_r is the *reduced side length,* while I and III give $h_1 = h\sqrt{1 + i_1}$, $h_3 = h\sqrt{1 + i_3}$, as

$$m = \tfrac{1}{6}\, p h^2 \qquad (1.112)$$

Further, if the quantity

$$\frac{b_r}{2} = \frac{b}{\sqrt{1 + i_1} + \sqrt{1 + i_3}} \qquad (1.113)$$

is introduced, II and IV both yield

$$m = \frac{p a_r^2}{24}\left(3 - 4\,\frac{h}{b_r}\right) \qquad (1.114)$$

It will now be seen that equations (1.112) and (1.114) have the same form as those to which the original equations are reduced when the slab is simply supported, i.e.:

> *The yield moment in a restrained slab is equal to the yield moment in a simply supported slab with the sides a_r and b_r.*

This elegant theorem was given by Ingerslev already in 1921. From (1.112) and (1.114), we find without difficulty:

$$2\,\frac{h}{a_r} = \sqrt{3 + \left(\frac{a_r}{b_r}\right)^2} - \frac{a_r}{b_r} \qquad (1.115)$$

$$m = \frac{p a_r^2}{24}\left[\sqrt{3 + \left(\frac{a_r}{b_r}\right)^2} - \frac{a_r}{b_r}\right]^2 \qquad (1.116)$$

The formulae only apply in the case of $h_1 + h_3 \leqslant b$, which can also be written $2h \leqslant b_r$, after which (1.115) gives $b_r \geqslant a_r$. The formulae must thus be used in such a way that b_r is greater than a_r.

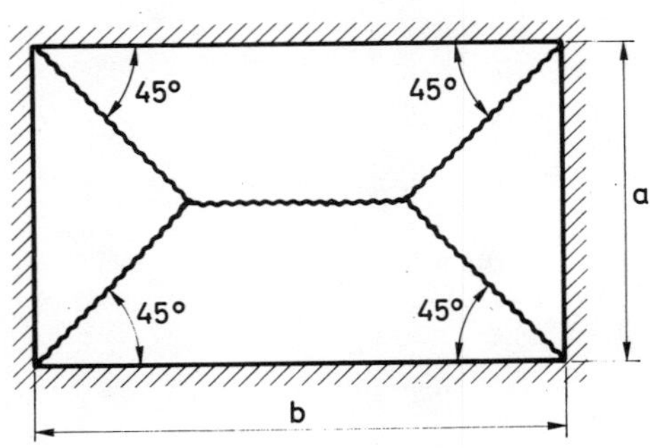

Fig. 1.112

For the simply supported slab, the work equation for the yield pattern shown in Fig. 1.112, in which all parts of the slab have the same rotation, gives

$$2ma + 2mb = 2 \times \frac{1}{2}\, pa \frac{a}{2}\,\frac{a}{6} + 2 \times \frac{1}{2}\, pb \frac{a}{2}\,\frac{a}{6} + 2 \times \frac{1}{2}\, p\,(b - a)\,\frac{a}{2}\,\frac{a}{3}$$

$$m = \frac{pa^2}{24}\,\frac{3b - a}{a + b} \qquad (1.116')$$

with a maximum deviation of about 2% from the

value of m determined by means of (1.116). Incidentally, the formula is identical with the formula known from the Danish standards of 1930, which was due to E. Suenson. By including the restraining moments, we get $m = m_o - m_i$, where m_o is the moment in a simply supported slab and m_i is the mean value

$$\frac{m_1'a + m_2'b + m_3'a + m_4'b}{2(a+b)} = m\,\frac{(i_1 + i_3)\,a + (i_2 + i_4)b}{2(a+b)}$$

The reactions $\bar{q}$ are uniformly distributed on all four sides and, in accordance with equation (109) in YLT, we get

$$\bar{q}_1 = \frac{2}{3}\,ph_1 = \frac{2}{3}\sqrt{1+i_1}\,ph;\ \bar{q}_3 = \tfrac{2}{3}\sqrt{1+i_3}\,ph \tag{1.117a}$$

$$\bar{q}_2 = \frac{1}{3}\,ph_2\left[2 + 2\left(1 - \frac{h_1 + h_3}{b}\right) - \left(1 - \frac{h_1 + h_3}{b}\right)^2\right]$$

$$= \frac{1}{6}\sqrt{1+i_2}\left[3 - \left(\frac{a_r}{b_r}\right)^2\left(2\frac{h}{a_r}\right)^2\right]pa_r \tag{1.117b}$$

$$\bar{q}_4 = \frac{1}{6}\sqrt{1+i_4}\left[3 - \left(\frac{a_r}{b_r}\right)^2\left(2\frac{h}{a_r}\right)^2\right]pa_r$$

since

$$h_2 = \frac{a_r}{2}\sqrt{1+i_2},\ h_4 = \frac{a_r}{2}\sqrt{1+i_4} \text{ and}$$

$$\frac{h_1 + h_3}{b} = \frac{2h}{b_r}$$

At the corners we get anchorage forces H(§ 8 in YLT), which, for example, at the corner formed by lines 1 and 2, will be

$$H_{1,2} = (m + m_1')\frac{h_2}{h_1} + (m + m_2')\frac{h_1}{h_2}$$

$$= m(1 + i_1)\frac{h_2}{h_1} + m(1 + i_2)\frac{h_1}{h_2}$$

and correspondingly. By substituting m from (1.112) in the first term and m from (1.114) in the second term, together with h_1 and h_2 from (1.111), we obtain the corner force

$$H_{1,2} = \frac{1}{3}\sqrt{(1+i_1)(1+i_2)}\left(1 - \frac{h}{b_r}\right)pa_r h \tag{1.118}$$

and correspondingly in the other corners when the appropriate values of i are substituted in (1.118).

For checking purposes, we know that the sum of all reactions must be pab, and we can easily see that this is the case.

With the knowledge we now have of moments and reactions and of the relationship between restrained slabs and simply supported slabs, we can easily formulate a simple approximate formula. Let us consider a simply supported slab and the beams supporting it. The beams parallel with a have the ultimate moment M_a, and the beams parallel with b, the ultimate moment M_b. Then, in the mid-sections through slab and beams, we have

$$2M_a + mb = \tfrac{1}{8}\,pa^2b;\ 2M_b + ma = \tfrac{1}{8}\,pab^2$$

We know that the reactions of the slab on the beams are uniformly distributed, so

$$M_a = \tfrac{1}{8}\,\bar{q}_a a^2;\ M_b = \tfrac{1}{8}\,\bar{q}_b b^2$$

which gives

$$\bar{q}_a = \frac{1}{2}\,pb - 4m\,\frac{b}{a^2};\ \bar{q}_b = \frac{1}{2}\,pa - 4m\,\frac{a}{b^2}$$

For reasons of symmetry, the corner forces are equal in magnitude, as explained in YLT, § 8.

$$H = m(\cot\alpha + \tan\alpha) \gtrsim 2m$$

since the minimum value of the parenthesis is 2 for $\alpha = 45°$ and does not vary much from 2 even when α deviates quite considerably from 45°.

Finally, an equation of vertical projection for the entire slab gives

$$4H + pab = 2a\bar{q}_a + 2b\bar{q}_b = 2pab - 8m\left(\frac{a}{b} + \frac{b}{a}\right)$$

When H is expressed by means of m, we get

$$m \leqslant \frac{pab}{8\left(1 + \frac{b}{a} + \frac{a}{b}\right)} \tag{1.116″}$$

which is a few per cent on the safe side when the equality sign is used. With this value of m, we get

$$\bar{q}_a = \bar{q}_b = \frac{1}{2}\,\frac{p(a+b)}{1 + \frac{a}{b} + \frac{b}{a}} = 4m\left(\frac{1}{a} + \frac{1}{b}\right) \tag{1.117}$$

It should be noted that a and b occur symmetrically in these formulae, which are therefore valid whether $a > b$ or $a < b$, and this is one of the advantages of this formula.

In the case of restrained slabs, a_r and b_r are substituted for a and b. A glance at the exact expressions in (1.117) and (1.118) will show that $\bar{q}_1$ is multiplied by $\sqrt{1+i_1}$, $\bar{q}_2$ by $\sqrt{1+i_2}$, etc. The corner forces are determined by means of

$$H_{1,2} = 2m\sqrt{1+i_1}\,\sqrt{1+i_2} \tag{1.119}$$

and correspondingly.

In calculating the beams supporting the slab, only the $\bar{q}$'s are used, while the corner forces H should be taken into account in the calculation of the columns as they reduce the reactions of the beams on the columns because they act in the opposite direction to $\bar{q}$.

The influence of the corner levers can be investigated as shown in YLT, page 99, but as they only entail a maximum increase in m of 8%, they are of no practical importance in these slabs.

1.12 Triangular load distribution

Applying the work equation for the yield pattern in Fig. 1.112, we get $\Sigma M_i + \Sigma M_u = 0$ since θ is identical for all parts of the slab. In this equation, M_i denotes the component in the direction of the axis of rotation of the resultant of the ultimate moments for one of the parts of the slab, and M_u denotes the moment of the external forces on this part of the slab about the axis of rotation, see also YLT, p. 63. Let us now imagine that the triangular loading is composed of a uniformly distributed load that results in the same total load as the triangular loading, i.e. that is equal to the mean load, and a positive triangular loading on one half of the slab, and a corresponding, negative loading on the other half. Two of the four parts of the slab in the yield pattern receive equal contributions from the positive and the negative triangular loading, so the total contribution only corresponds to the uniformly distributed part of the load. One of the remaining two parts of the slab receives a contribution from the positive triangular loading, and the other, a corresponding contribution from the negative loading. Therefore, in ΣM_u, we only get the contributions from the uniformly distributed part of the load, i.e. the ultimate moment will be the same as in a slab with the same total load, uniformly distributed. This result corresponds to the conditions in a one-way slab with triangular loading.

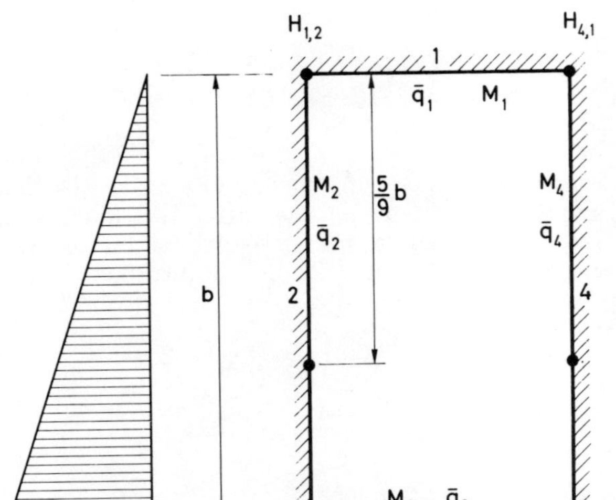

Fig. 1.12

The exact calculation shows that this is actually an excellent approximation and that restrained slabs can be calculated as simply supported slabs with the reduced spans a_r and b_r in accordance with (1.111) and (1.113).

Formula (1.116″) and expressions for the reactions can be derived in a similar manner.

The simply supported slab has the reactions $\bar{q}_1$ and $\bar{q}_3$, uniformly distributed, and the equal reactions $\bar{q}_2$ and $\bar{q}_4$, which are not uniformly distributed. If, however, $\bar{q}_2$ and $\bar{q}_4$ are assumed to vary linearly, the beam moments can be reckoned to be $M_2 = M_4 \sim \frac{1}{8}\bar{q}_2 b^2$, where $\bar{q}_2$ is the mean value. In a section through slab and beams we get, as previously, the following equations, in which P is the total load:

$$M_1 + M_3 + mb = \frac{P}{8} a$$

$$2M_2 + ma \sim \frac{P}{8} b$$

Further,

$$(\bar{q}_1 + \bar{q}_3)\, a + 2\bar{q}_2 b = P + \Sigma H$$

and, as $M \simeq \frac{1}{8}\bar{q}l^2$ for all four beams, we get, as before,

$$P = 8m\left(\frac{a}{b} + \frac{b}{a}\right) + \Sigma H$$

where all $H \geqslant 2m$, i.e. $\Sigma H \geqslant 8m$, which immediately gives (1.116″) and

$$2\bar{q}_2 = \bar{q}_1 + \bar{q}_3 = 8m\left(\frac{1}{a} + \frac{1}{b}\right) \qquad (1.121)$$

In this way we can thus determine only the sum of $\bar{q}_1$ and $\bar{q}_3$.

The exact calculation now shows that the resultant of the reactions $\bar{q}_2$ on side 2 lies at a distance of 0·54-0·58b from side 1, where p = 0, corresponding to a variation of b/a from 0 to 4. This distance can thus with good approximation be reckoned as $\frac{5}{9}$b. As $\bar{q}_2$ is assumed to vary linearly from $\bar{q}_1$ to $\bar{q}_3$, this distance depends upon the ratio between $\bar{q}_1$ and $\bar{q}_3$, and when $\bar{q}_3 = 2\bar{q}_1$, we get precisely $\frac{5}{9}$b. With this value we then have

$$\bar{q}_1 = \frac{2}{3}\bar{q}_2;\ \bar{q}_3 = \frac{4}{3}\bar{q}_2;\ \bar{q}_2 = 4m\left(\frac{1}{a} + \frac{1}{b}\right) \qquad (1.122)$$

This distribution of the reactions corresponds to the conditions in the one-way slab with triangular loading.

In the case of restrained slabs, the reduced spans a_r and b_r are substituted, and $\bar{q}_1$ is multiplied by

$\sqrt{1 + i_1}$, etc., $H_{1,2}$ by $\sqrt{1 + i_1}\,\sqrt{1 + i_2}$, i.e. we get

$$\bar{q}_1 = \frac{8}{3}\, m\left(\frac{1}{a_r} + \frac{1}{b_r}\right)\sqrt{1 + i_1}$$

$$\bar{q}_2 = 4m\left(\frac{1}{a_r} + \frac{1}{b_r}\right)\sqrt{1 + i_2}$$

$$\bar{q}_3 = \frac{16}{3}\, m\left(\frac{1}{a_r} + \frac{1}{b_r}\right)\sqrt{1 + i_3} \qquad (1.123)$$

$$\bar{q}_4 = 4m\left(\frac{1}{a_r} + \frac{1}{b_r}\right)\sqrt{1 + i_4}$$

$$H_{1,2} = 2m\sqrt{1 + i_1}\,\sqrt{1 + i_2}$$

$$H_{2,3} = 2m\sqrt{1 + i_2}\,\sqrt{1 + i_3}$$

etc., and

$$m = \frac{\tfrac{1}{2} p a_r b_r}{8\left(1 + \dfrac{a_r}{b_r} + \dfrac{b_r}{a_r}\right)}$$

1.13 Line loads

As, in practice, a line load $\bar{p}$ is caused by light partition walls, we will only treat it here as a supplementary load to a uniformly distributed load. The following formulae are therefore not applicable in cases in which the line load is predominant, as this would lead to curved yield lines (YLT, example 30). As partition walls are likely to be pulled down and others built, they must be considered as a live load. Therefore, in the following, the moment is determined for the most dangerous position of the line load. Let us then consider $\bar{p}_a$ parallel with side a and acting on the middle part of the slab, and $\bar{p}_b$ parallel with side

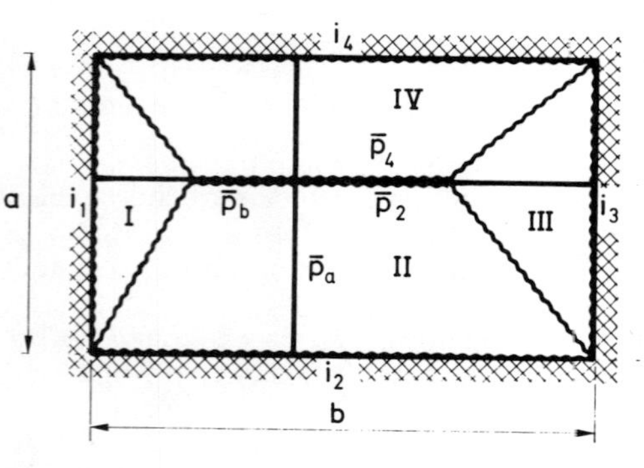

Fig. 1.13

b and acting in the yield line (see Fig. 1.13). The latter is distributed with $\bar{p}_2$ to slab part II and $\bar{p}_4$ to slab part IV, where $\bar{p}_2 + \bar{p}_4 = \bar{p}_b$. The equilibrium equations used earlier then become

$$m(1 + i_1) = \frac{1}{6}\, p h_1^2\left(1 + 3\frac{\bar{p}_b}{pa}\right)$$

$$m(1 + i_3) = \frac{1}{6}\, p h_3^2\left(1 + 3\frac{\bar{p}_b}{pa}\right) \qquad (1.131)$$

and

$$m(1 + i_2) = \frac{1}{6} p h_2^2\left(3 - \frac{2}{b}(h_1 + h_3) + 3\frac{\bar{p}_a}{pb}\right) + \bar{p}_2 h_2\left(1 - \frac{h_1 + h_3}{b}\right)$$

$$(1.132)$$

$$m(1 + i_4) = \frac{1}{6} p h_4^2\left(3 - \frac{2}{b}(h_1 + h_3) + 3\frac{\bar{p}_a}{pb}\right) + \bar{p}_4 h_4\left(1 - \frac{h_1 + h_3}{b}\right)$$

From the last two we get

$$m\left(\frac{1 + i_2}{h_2} + \frac{1 + i_4}{h_4}\right) = \frac{1}{6} pa\left[3\left(1 + \frac{\bar{p}_a}{pb}\right) - \frac{2}{b}(h_1 + h_3)\right] + \bar{p}_b\left(1 - \frac{h_1 + h_3}{b}\right) \qquad (1.133)$$

As $\partial m/\partial h_2 = 0$, $\bar{p}_b$ being assumed to be placed in the most dangerous position, we immediately find that, since $h_2 + h_4 = a$,

$$\frac{h_2}{\sqrt{1 + i_2}} = \frac{h_4}{\sqrt{1 + i_4}} = \frac{a}{\sqrt{1 + i_2} + \sqrt{1 + i_4}} = \frac{1}{2} a_r \qquad (1.111)$$

It will be seen from (1.131) that

$$\frac{h_1}{\sqrt{1 + i_1}} = \frac{h_3}{\sqrt{1 + i_3}} = h$$

Substituting b_r as earlier (1.113), we get

$$m = \tfrac{1}{6}\, p h^2 (1 + 3\beta) \qquad (1.134)$$

and

$$m = \frac{1}{24} p a_r^2\left[3(1 + \alpha + 2\beta) - 4(1 + 3\beta)\frac{h}{b_r}\right] \qquad (1.135)$$

where $\alpha = \bar{p}_a/pb = \bar{p}_a a/pab$ and $\beta = \bar{p}_b/pa = \bar{p}_b b/pab$ are the ratios between the entire line load in the a and the b direction, respectively, and the entire uniformly distributed load. The introduction here of

$$p' = p(1 + \alpha + 2\beta) = p + \frac{\bar{p}_a}{b} + 2\frac{\bar{p}_b}{a} \qquad (1.136)$$

$$a' = a_r,$$

$$b' = b_r\sqrt{\frac{1 + \alpha + 2\beta}{1 + 3\beta}},\quad h' = h\sqrt{\frac{1 + 3\beta}{1 + \alpha + 2\beta}} \qquad (1.137)$$

results in equations (1.112) and (1.114) for uniformly distributed load only. This means that the slab with p, $\bar{p}_a$ and $\bar{p}_b$ has the same ultimate moment m as the slab with only p' and the sides a' and b'. If, in particular, the two total line loads are of equal magnitude, we get $\alpha = \beta$, $a' = a_r$ and $b' = b_r$, so we have the same slab, but with $p' = p(1 + 3\alpha)$.

Normally, $\bar{p}_a = \bar{p}_b$, and then $\alpha = \beta$ for $a_r = b_r$. The supplementary load will then be $3(\bar{p}/a_r)$. If the wall has a height h_v and weight V per m^2, then $\bar{p} = Vh_v$ and the supplementary load is $3(h_v/a_r)V$. For each wall we get $(3h_v/2a_r)V$. This value does not exceed V as long as $h_v \leqslant \frac{2}{3}a_r$.

1.14 Point loads (wheel loads)

As shown in YLT, page 82, the yield pattern in the case of a point load P is a circle and $P = 2\pi(m + m')$. Therefore, for any position of P in a simply supported, rectangular slab, we get

$$m = P/2\pi \qquad (1.141)$$

If all four sides are restrained with identical $m' = im$, we find that

$$m = P/2(1 + i)\pi \qquad (1.142)$$

If only three sides are restrained, the following will apply at the simply supported edge, in accordance with YLT, page 83, formula (45):

$$m = P\Big/2\left[\sqrt{i} + (1 + i)\left(\frac{\pi}{2} + \text{arccot}\,\sqrt{i}\right)\right] \qquad (1.143)$$

In particular, $i = 1$ yields

$$m = P/(2 + 3\pi) = P/11{\cdot}4$$

If two opposite sides are restrained and the load is applied half-way between them, we obtain, in accordance with YLT, example 27, page 111,

$$m = P/4[\sqrt{i} + (1 + i)\,\text{arccot}\,\sqrt{i}] \qquad (1.144)$$

and, in particular, for $i = 1$,

$$m = P/(4 + 2\pi) = P/10{\cdot}3$$

However, the yield pattern can only be formed if the length of the simply supported sides is greater than or equal to $\sqrt{1 + i}$ times that of the restrained sides. If this is not so, we get a lower value of m.

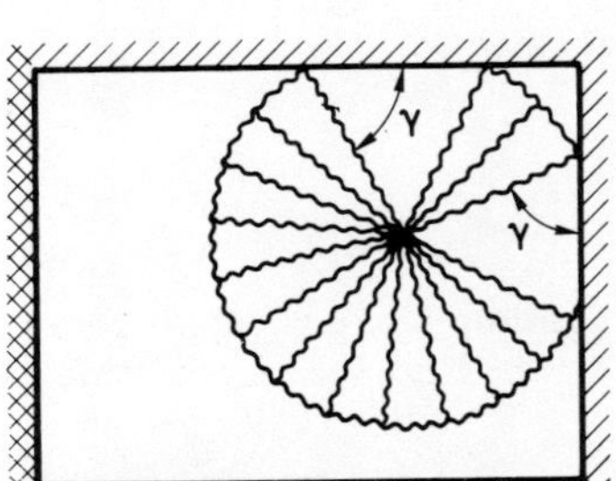

Fig. 1.141

If two adjoining sides are restrained, the yield pattern shown in Fig. 1.141 again results in (1.144).

With one side restrained, the yield pattern shown in Fig. 1.142 gives

$$P = (m + m')(6\gamma - \pi) + m \times 6 \cot\gamma, \ \cot\gamma = \sqrt{i}$$

$$m = P/6\left[\sqrt{i} + (1 + i)\left(\text{arccot}\,\sqrt{i} - \frac{\pi}{6}\right)\right] \qquad (1.145)$$

and, in particular, for $i = 1$,

$$m = P/(6 + \pi) = P/9{\cdot}14$$

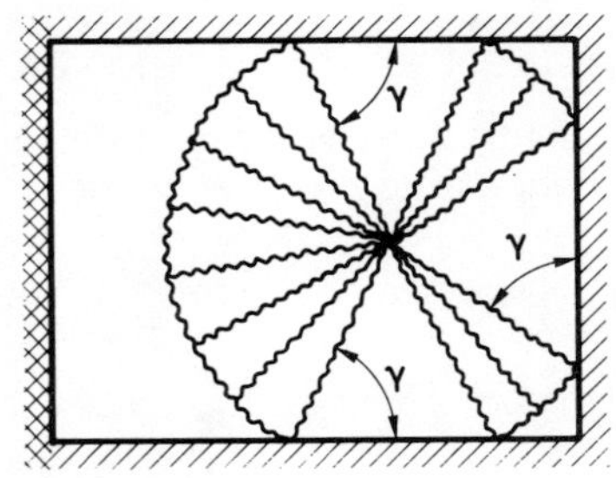

Fig. 1.142

For this yield pattern to be formed, the simply supported side at the restraint must be at least $\frac{1}{2}(1 + \sqrt{1 + i})$ times the restrained side; when this is not the case, more favourable conditions prevail.

The formulae derived (1.141 to 1.145) can be approximated with sufficient accuracy as follows, as long as $i \leqslant 1$:

$$m = P/[6{\cdot}3 + (1{\cdot}8 + 1{\cdot}1n)i] \qquad (1.14)$$

where n is the number of restrained sides. It should be noted that the top reinforcement must be carried through the entire slab as, otherwise, the circle corresponding to $m = P/2\pi$ and of appropriate radius may form.

In general, wheel loads are assumed to be distributed over a small rectangle, and formulae for this case are specified in Bygningsstatiske Meddelelser, Vol. IV, page 77, so it will suffice here to intimate their derivation in brief. As the moment in a beam with span a and the load P distributed over a short length l is $\frac{1}{8}P\,(2a - l)$, we can expect the following to apply in a rectangular slab with P distributed over the rectangle with the sides k and l:

$$m_P = \frac{P}{24}\,\frac{2a - k}{a}\,\frac{2b - l}{b} \qquad (1.146)$$

provided the lengths of sides a and b do not differ too much. This assumption is confirmed by detailed investigations in the paper cited.

If there is a possibility of two equal wheel loads P acting on the slab, the work equation for the yield pattern shown in Fig. 1.143 in a square slab gives

$$m\,\frac{a}{x} + m\,\frac{a}{a - c - x} + 2ma\,\frac{2}{a} = 2P \times 1$$

which gives maximum m for $x = \frac{1}{2}(a - c)$, i.e. the loads must be equidistant from the middle. We then get

$$m_{2P} = \frac{P}{2}\,\frac{a-c}{2a-c} = 4m_P\,\frac{a-c}{2a-c} \qquad (1.147)$$

since $c = 0$ yields $m_{2P} = 2m_P$. When one of the loads acts partly outside the slab, we get, by interpolation,

$$m_{P+P'} = 2\left(1 - \frac{c}{a+k}\right)m_P \qquad (1.148)$$

In the case of restrained slabs, we assume that $m = m_o - m_i$, where m_i is the mean value of m_1', m_2', m_3', m_4'.

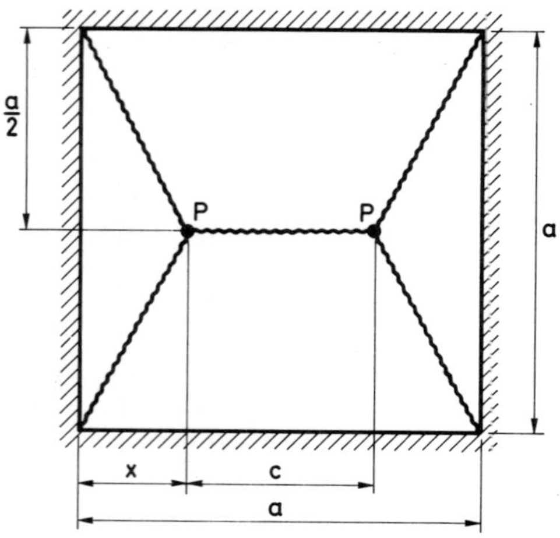

Fig. 1.143

The formulae developed are too conservative when there is a considerable difference in the sides. However, as long as the ratio between the longest and the shortest side does not exceed 3/2, the formulae can be used to advantage.

1.15 Slabs with holes

In the case of a central hole, the yield pattern will be as shown in Fig. 1.151. With a uniformly distributed

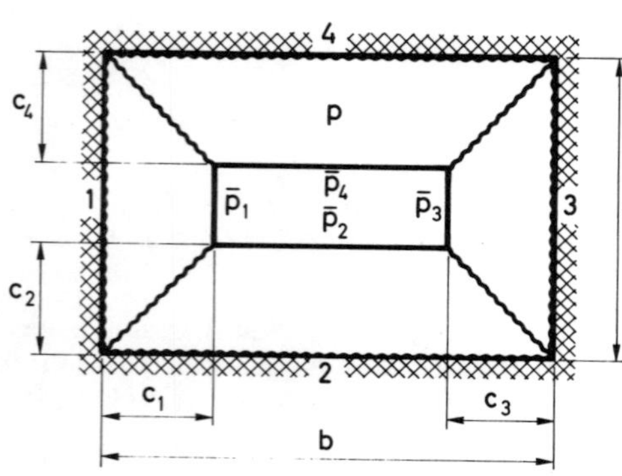

Fig. 1.151

load p and a line load $\bar{p}_m$ ($m = 1, 2, 3, 4$) acting along the edge of the hole, we find from the work equation when there is unit deflexion at the hole, whereby the rotations are $1/c_1$, $1/c_2$, etc., that

$$\left[m_1'a + m(c_2 + c_4)\right]\frac{1}{c_1} + \left[m_2'b + m(c_1 + c_3)\right]\frac{1}{c_2}$$

$$+\left[m_3'a + m(c_2 + c_4)\right]\frac{1}{c_3} + \left[m_4'b + m(c_1 + c_3)\right]\frac{1}{c_4}$$

$$= \left[\frac{1}{2}pac_1^2 - \frac{1}{3}p(c_2 + c_4)c_1^2\right]\frac{1}{c_1} + \bar{p}_1(a - c_2 - c_4)$$

$$+\left[\frac{1}{2}pbc_2^2 - \frac{1}{3}p(c_1 + c_3)c_2^2\right]\frac{1}{c_2}$$

$$+\bar{p}_2(b - c_1 - c_3) + \text{etc.}$$

Rearranging the equations and introducing the i's, we obtain

$$m\left[\left(\frac{i_1}{c_1} + \frac{i_3}{c_3}\right)a + \left(\frac{i_2}{c_2} + \frac{i_4}{c_4}\right)b + (c_1 + c_3)\left(\frac{1}{c_2} + \frac{1}{c_4}\right)\right.$$

$$\left. + (c_2 + c_4)\left(\frac{1}{c_1} + \frac{1}{c_3}\right)\right]$$

$$= \frac{1}{2}pa(c_1 + c_3) + \frac{1}{2}pb(c_2 + c_4)$$

$$-\frac{1}{3}p(c_1 + c_3)(c_2 + c_4) + (\bar{p}_1 + \bar{p}_3)(a - c_2 - c_4)$$

$$+ (\bar{p}_2 + \bar{p}_4)(b - c_1 - c_3)$$

$$= \frac{1}{2}P + \bar{P} - \frac{1}{6}p(c_1 + c_3)(c_2 + c_4) \qquad (1.151)$$

where P is the total uniformly distributed load and $\bar{P}$ is the total line load, i.e.

$$P = p[a(c_1 + c_3) + b(c_2 + c_4) - (c_1 + c_3)(c_2 + c_4)]$$

$$\bar{P} = (\bar{p}_1 + \bar{p}_3)(a - c_2 - c_4) + (\bar{p}_2 + \bar{p}_4)(b - c_1 - c_3)$$

In the particular case in which all $i = 0$ and $c_1 = c_2 = c_3 = c_4 = c$, we get

$$m = \frac{P}{16} + \frac{\bar{P}}{8} - \frac{pc^2}{12} \qquad (1.152)$$

For all $i = 0$, $c_4 = 0$, and $c_1 = c_2 = c_3 = c$, we get the yield pattern shown in Fig. 1.152, for which the work equation gives

$$2mc\,\frac{1}{x} + 2m(c + x)\,\frac{1}{c} + 2mc\,\frac{1}{c}$$

$$= 2\left[\frac{1}{6}pcx^2 + \frac{1}{2}\bar{p}x^2\right]\frac{1}{x} + 2\left[\frac{1}{2}pac^2 - \frac{1}{3}p(x + c)c^2\right.$$

$$+ \bar{p}(a - x - c)c\Big]\frac{1}{c}$$

$$+\left[\frac{1}{2}\,pbc^2 - \frac{2}{3}\,pc^3 + \bar{p}(b - 2c)c\right]\frac{1}{c}$$

or

$$2m\left(2 + \frac{x}{c} + \frac{c}{x}\right) = \frac{1}{2}\,P + \bar{P} - \frac{1}{3}\,pc^2 - x\left(\frac{1}{3}pc + \bar{p}\right)$$

where

$$P = pc(2a + b - 2c)$$

$$\bar{P} = \bar{p}\,(2a + b - 4c)$$

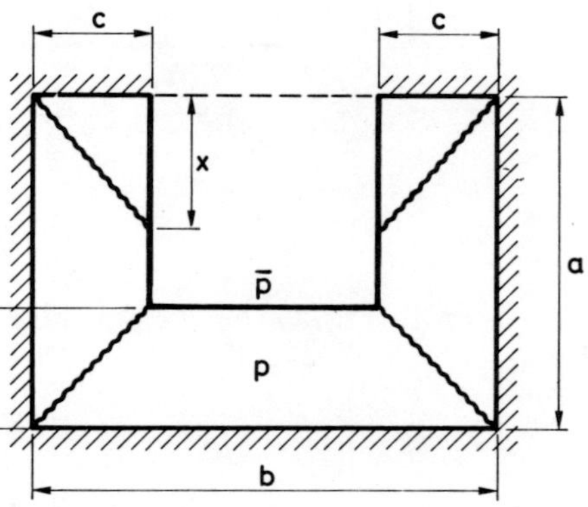

Fig. 1.152

It will immediately be seen that

$$\frac{P}{16} + \frac{\bar{P}}{8} - \frac{pc^2}{12} - \frac{\bar{p}c}{8} < m < \frac{P}{16} + \frac{\bar{P}}{8} - \frac{pc^2}{24} \qquad (1.153)$$

where we get the lower bound for $x = c$ and the upper bound by omitting the last negative term on the right-hand side and using the minimum value of the left-hand side.

By putting $\partial m/\partial x = 0$, we could find the exact value of m, but this will seldom be necessary. The correct value of x is slightly lower than c.

When c_3 is also equal to zero, the yield pattern shown in Fig. 1.153 gives the following equation (the fact that x must be the same at both ends is obvious from the moment equation for the triangular parts of the slab):

$$2mc\,\frac{1}{x} + 2mx\,\frac{1}{c} + 2mc\,\frac{1}{c} = 2\left[\frac{1}{6}\,pcx^2 + \frac{1}{2}\,\bar{p}x^2\right]\frac{1}{x}$$

$$+\left[\frac{1}{2}\,pac^2 - \frac{1}{3}\,p(x + c)c^2\right]\frac{1}{c} + \left[\frac{1}{2}\,pbc^2\right.$$

$$\left. - \frac{1}{3}\,p(x + c)c^2\right]\frac{1}{c} + \bar{p}\,(a - x - c + b - x - c)$$

or

$$2m\left(1 + \frac{c}{x} + \frac{x}{c}\right) = \frac{1}{2}\,P + \bar{P} - \frac{1}{6}\,pc^2 - x\left(\frac{1}{3}\,pc + \bar{p}\right)$$

where

$$P = pc(a + b - c)$$

$$\bar{P} = \bar{p}(a + b - 2c)$$

Thence, as above, we get

$$\frac{P}{12} + \frac{\bar{P}}{6} - \frac{pc^2}{12} - \frac{\bar{p}c}{6} < m < \frac{P}{12} + \frac{\bar{P}}{6} - \frac{pc^2}{36} \qquad (1.154)$$

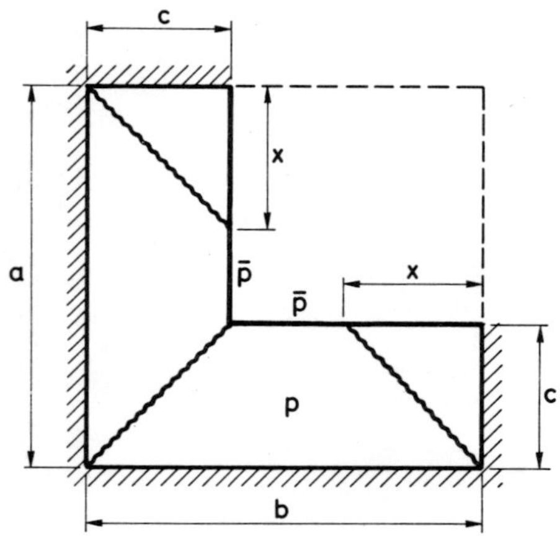

Fig. 1.153

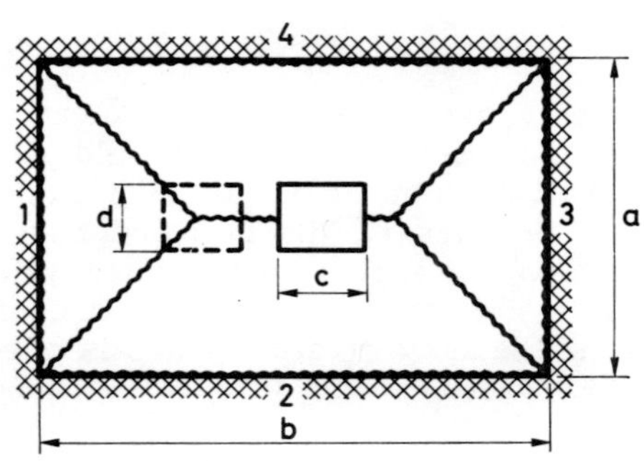

Fig. 1.154

When c is small in relation to a or b, more dangerous yield patterns arise, and these will be dealt with in section 1.21.

In the case of small holes we can, as an approximation, reckon on loading over the hole, whereby the contribution from the external forces remains unaltered. As far as the internal forces are concerned (Fig. 1.154), we get $m(1 + i_2)b$ changing to $m(b - c) + i_2mb = m[1 + i_2 - (c/b)]b$ and correspondingly on the opposite side. In other words, the degrees of restraint i_2 and i_4 are reduced by c/b. In the case of simple supports we thus get, purely formally, negative degrees of restraint $-(c/b)$. The "reduced" span a_r will then be greater than a. As $1 - (c/b)$ is always positive, the square root will always be real. If the hole lies slightly eccentrically, we get the further reduction d/a in the degree of restraint at one of the sides a. If the slab is nearly square, we get a reduction in the degree of restraint on all four sides.

In the case of medium-sized eccentric holes, a large number of different yield patterns arise, so these cases do not lend themselves to the establishment of ready-made formulae. However, m can be determined quite quickly by means of the work equation and a suitably estimated yield pattern.

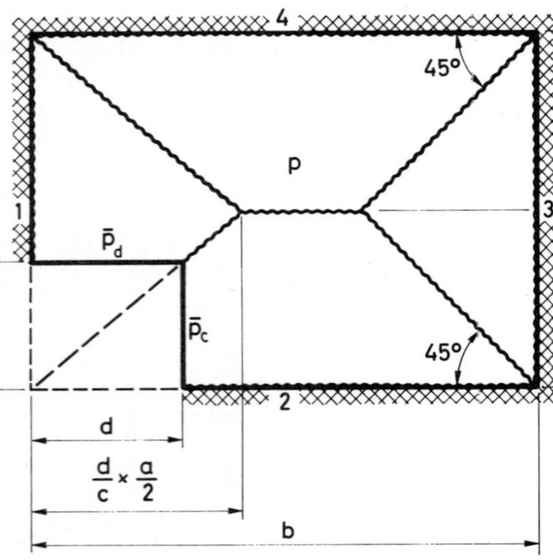

Fig. 1.155

We shall demonstrate this fact in two common cases. When there is a hole in one corner, the yield pattern shown in Fig. 1.155 can be used, with a yield line passing through the corner of the hole and using the bisectors of the corner angles for the other two yield lines through the corners. Three of the parts of the slab thus get the same rotation, which is fixed at 1. The fourth part then gets $\theta = c/d$. We then have

$$m(1 + i_1)(a - c)\frac{c}{d} + m(1 + i_2)(b - d) + m(1 + i_3)a$$

$$+ m(1 + i_4)\,b = \left[\frac{1}{6}pa\frac{d^2}{c^2}\frac{a^2}{4} - \frac{1}{6}pcd^2 + \frac{1}{2}\bar{p}_d d^2\right]\frac{c}{d}$$

$$+ \frac{1}{6}p\frac{d}{c}\frac{a}{2}\frac{a^2}{4} - \frac{1}{6}pdc^2 + \frac{1}{2}p\left(b - \frac{d}{c}\frac{a}{2}\right)\frac{a^2}{4} - \frac{1}{3}p\frac{a}{2}\frac{a^2}{4}$$

$$+ \frac{1}{6}p\frac{a}{2}\frac{a^2}{4} + \frac{1}{2}pb\frac{a^4}{4} - \frac{1}{3}p\left(\frac{a}{2} + \frac{d}{c}\frac{a}{2}\right)\frac{a^2}{4} + \frac{1}{2}\bar{p}_c c^2$$

$$= \frac{pa^2}{24}(6b - a) + \frac{1}{2}\bar{P}c - \frac{pd}{24c}(a^3 + 8c^3)$$

$$= \frac{pa^2}{24}\left[6b - \left(1 + \frac{d}{c}\right)a\right] - \frac{1}{3}pc^2d + \frac{1}{2}\bar{P}c \qquad (1.155)$$

where $\bar{P} = \bar{p}_c c + \bar{p}_d d$.

In the special case of a square hole (c = d), we get

$$m[(1 + i_1)(a - c) + (1 + i_2)(b - c) + (1 + i_3)a$$

$$+ (1 + i_4)b] = \frac{pa^2}{12}(3b - a) - \frac{1}{3}pc^3 + \frac{1}{2}\bar{P}c$$

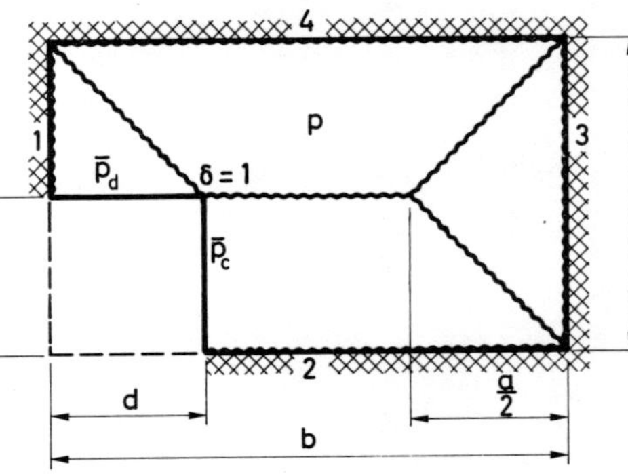

Fig. 1.156

In the case shown in Fig. 1.156, we get the rotations $\theta_1 = 1/d$, $\theta_2 = 1/c$, $\theta_3 = 2/a$, $\theta_4 = 1/(a - c)$ and the work equation

$$m(a - c)\frac{1 + i_1}{d} + m(b - d)\frac{1 + i_2}{c} + ma\frac{2}{a}(1 + i_3)$$

$$+ mb\frac{1 + i_4}{a - c} = \left[\frac{1}{6}p(a - c)d^2 + \frac{1}{2}\bar{p}_d d^2\right]\frac{1}{d}$$

$$+ \left[\frac{1}{2}p(b - d)c^2 - \frac{1}{3}p\frac{a}{2}c^2 + \frac{1}{2}\bar{p}_c c^2\right]\frac{1}{c} + \frac{1}{6}pa\frac{a^2}{4}$$

$$\times \frac{2}{a} + \left[\frac{1}{2}pb(a - c)^2 - \frac{1}{3}p\left(d + \frac{a}{2}\right)(a - c)^2\right]\frac{1}{a - c}$$

$$= \frac{1}{12}pa(6b - a) - \frac{1}{6}pd(a + 2c) + \frac{1}{2}\bar{P}. \qquad (1.156)$$

The formula that gives maximum m should be used.

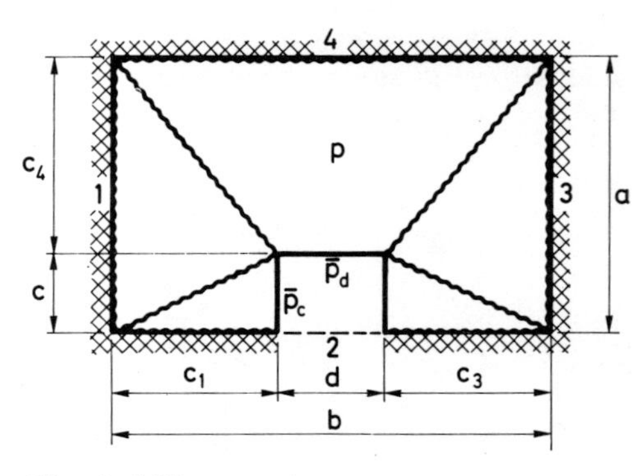

Fig. 1.157

When there is a hole at the edge, as shown in Fig. 1.157, we get the rotations $\theta_1 = 1/c_1$, $\theta_2 = 1/c$, etc., and the work equation

$$ma\left(\frac{1+i_1}{c_1}+\frac{1+i_3}{c_3}\right)+m(c_1+c_3)\left(\frac{1+i_2}{c}+\frac{1+i_4}{c_4}\right)$$

$$=\bar{p}_d d+2\bar{p}_c\frac{c}{2}+\frac{1}{6}pac_1{}^2\frac{1}{c_1}$$

$$+\frac{1}{6}pc_1c^2\frac{1}{c}+\frac{1}{6}pc_3c^2\frac{1}{c}+\frac{1}{6}pac_3{}^2\frac{1}{c_3}$$

$$+\left[\frac{1}{2}pbc_4{}^2-\frac{1}{3}p(c_1+c_3)c_4{}^2\right]\frac{1}{c_4}$$

$$=\frac{1}{6}pa(2b+d)-\frac{1}{2}pdc+\bar{p}_c c+\bar{p}_d d \qquad (1.157)$$

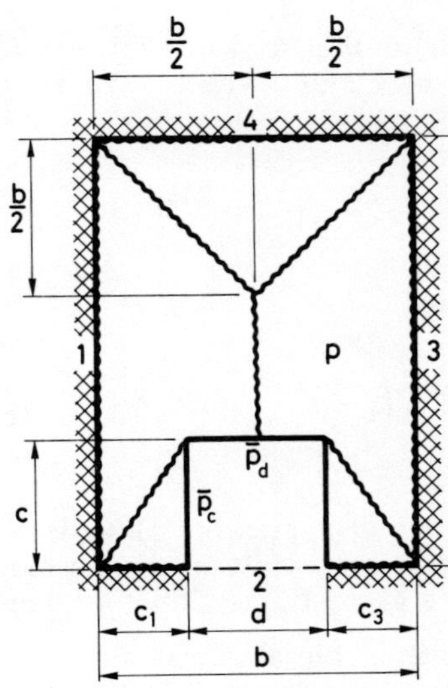

Fig. 1.158

The yield pattern shown in Fig. 1.158 results in the same rotation θ for three of the parts, and making this rotation equal to 1, we get $\theta = b/2c$ for the fourth part. The work equation then yields

$$ma(1+i_1+1+i_3)+mb(1+i_4)$$

$$+m(c_1+c_3)(1+i_2)\frac{b}{2c}=\bar{p}_d d\frac{b}{2}+2\bar{p}_c c\frac{b}{4}$$

$$+2\times\frac{1}{2}p(a-c)\frac{b^2}{4}-2\times\frac{1}{3}p\frac{b}{2}\frac{b^2}{4}+\frac{1}{6}pb\frac{b^2}{4}$$

$$+\frac{1}{6}pc(c_1{}^2+c_3{}^2)+\frac{1}{6}p(c_1+c_3)c^2\frac{b}{2c}$$

or

$$m\left[\frac{2a}{b}(2+i_1+i_3)+2(1+i_4)+(1+i_2)\frac{b-d}{c}\right]$$

$$=\frac{pb}{12}(6a-b-6c)+\frac{1}{6}p\frac{c}{b}[2c_1{}^2+2c_3{}^2$$

$$+b(b-a)]+\bar{p}_c c+\bar{p}_d d \qquad (1.158)$$

As previously, it is the formula that gives maximum m that should be used. Formulae (1.155) to (1.158) are lower bounds for m, as they are derived by means of the work equation. If the load on the edge of the hole does not greatly exceed the load pcd corresponding to the area of the hole, an upper bound for m can be obtained by reducing the degrees of restraint along the sides a and b by c/a and d/b, respectively, and reckoning the slab to have no hole, i.e. applying (1.116″) with the transformed spans

$$a_t=\frac{2a}{\sqrt{1+i_2-\frac{c}{a}}+\sqrt{1+i_4-\frac{c}{a}}}$$

and

$$b_t=\frac{2b}{\sqrt{1+i_1-\frac{d}{b}}+\sqrt{1+i_3-\frac{d}{b}}} \qquad (1.159)$$

Similar expressions can be used for slabs with more than one hole. In these cases it is also necessary to bear in mind the possibility of local failure between the holes.

1.16 Extent of reinforcement at the restrained edges

When the reinforcement is carried a distance $\alpha_1 b$ into the panel at the restrained edge 1 and a distance $\alpha_2 a$ into the panel at edge 2, etc., and the panel is

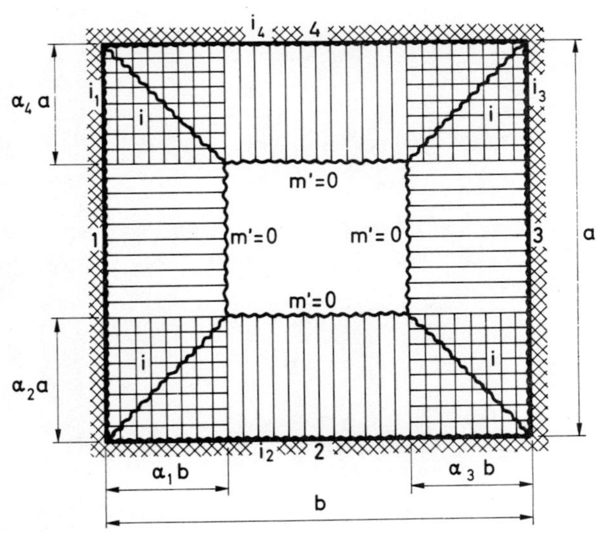

Fig. 1.16

only subjected to the dead load g, while the adjacent panels are fully loaded, the yield pattern shown, with negative moments, may arise. The middle part of the panel is lifted up. The ultimate moments are zero at the boundary of the top reinforcement and im along the yield lines to the corners. The restraining moments are $i_1m, i_2m, \ldots$ etc., where the i's are either equal to i or zero. There are thus also bent-up bars in the case of simple supports.

When the middle of the slab is lifted 1, the work equation gives

$$[i_1 ma - (\alpha_2 + \alpha_4) ima]\frac{1}{\alpha_1 b}$$

$$+ [i_2 mb - (\alpha_1 + \alpha_3) imb]\frac{1}{\alpha_2 a} + \ldots \text{etc.}$$

$$= ga(1 - \alpha_2 - \alpha_4)b(1 - \alpha_1 - \alpha_3) \times 1$$

$$+ \left[\frac{1}{2} ga\alpha_1{}^2 b^2 - \frac{1}{3} ga(\alpha_2 + \alpha_4)\alpha_1{}^2 b^2\right]\frac{1}{\alpha_1 b} + \ldots \text{etc.}$$

or

$$m\frac{a}{b}\left[\frac{i_1}{\alpha_1} + \frac{i_3}{\alpha_3} - (\alpha_2 + \alpha_4)\left(\frac{1}{\alpha_1} + \frac{1}{\alpha_3}\right) i\right]$$

$$+ m\frac{b}{a}\left[\frac{i_2}{\alpha_2} + \frac{i_4}{\alpha_4} - (\alpha_1 + \alpha_3)\left(\frac{1}{\alpha_2} + \frac{1}{\alpha_4}\right) i\right]$$

$$= gab\left[1 - \frac{1}{2}(\alpha_1 + \alpha_2 + \alpha_3 + \alpha_4)\right.$$

$$\left. + \frac{1}{3}(\alpha_1 + \alpha_3)(\alpha_2 + \alpha_4)\right] \qquad (1.161)$$

In the special case in which all α's are identical, we get

$$m\left[\frac{a}{b}(i_1 + i_3 - 4\alpha i) + \frac{b}{a}(i_2 + i_4 - 4\alpha i)\right]$$

$$= \alpha\left(1 - 2\alpha + \frac{4}{3}\alpha^2\right) gab \qquad (1.162)$$

The reinforcement can also be carried in an equal distance on all sides, i.e. $\alpha_1 b = \alpha_3 b = \alpha_2 a = \alpha_4 a = \alpha a$, whence

$$m[i_1 + i_2 + i_3 + i_4 - 8\alpha i]$$

$$= \alpha\left[1 - \left(1 + \frac{a}{b}\right)\alpha + \frac{4}{3}\frac{a}{b}\alpha^2\right] gab \qquad (1.163)$$

For a square slab, with n restraints and

$$m = m_o - m_i = \frac{qa^2}{24} - \frac{m}{4}(i_1 + i_2 + i_3 + i_4)$$

we get, when $i = 1$:

$$\frac{q}{g} = 6\frac{4 + n}{n - 8\alpha}\alpha\left(1 - 2\alpha + \frac{4}{3}\alpha^2\right) \qquad (1.164)$$

which yields the following values for q/g:

α	n = 4	3	2	1
0·1	1·22	1·55	2·44	12·2
0·15	1·88	2·56	4·92	—
0·20	2·61	3·92	11·8	—
0·25	3·50	6·11	∞	—
0·30	4·69	12·5	—	—
0·35	6·48	34·0	—	—
0·40	9·94	—	—	—
0·45	20·0	—	—	—
0.50	∞	—	—	—

It will be seen from this that $\alpha = 0{\cdot}2$ will usually suffice. For rectangular slabs with $b = 2a$, we get

$$m = \frac{qab}{28} - \frac{m}{6}(i_1 + 2i_2 + i_3 + 2i_4)$$

and then (1.163) gives

$$\frac{q}{g} = \frac{14}{3}\frac{6 + n_a + 2n_b}{n_a + n_b - 8\alpha}\alpha\left(1 - \frac{3}{2}\alpha + \frac{2}{3}\alpha^2\right) \qquad (1.165)$$

where n_a is the number of restraints along the sides a, and n_b the number of restraints along the sides b. The values of q/g are given in the Table on page 15.

In the above investigation, the slab was only subjected to its dead load g. When the slab is totally loaded with q, the top reinforcement has to be carried so far in that the moment in the part of the middle panel without top reinforcement does not exceed the moment of the whole slab. As the middle is calculated as a simply supported slab with the spans $a(1 - \alpha_2 - \alpha_4)$ and $b(1 - \alpha_1 - \alpha_3)$, and the slab is calculated as a simply supported slab with the spans a_r and b_r, it follows that

$$a(1 - \alpha_2 - \alpha_4) \leq a_r, \quad b(1 - \alpha_1 - \alpha_3) \leq b_r \qquad (1.166)$$

With $i_1 = i_3 = 1$, we get $b_r = 0{\cdot}7b$ and $\alpha_1 = \alpha_3 = 0{\cdot}15$. With $i_1 = 1$ and $i_3 = 0$, we get $b_r = 0{\cdot}83b$, $\alpha_1 = 0{\cdot}17$ and $\alpha_3 = 0$.

The results are even more favourable than in the case of the square slab.

1.17 Deflexions

Two-way slabs must be regarded as a rather flexible type of structure, and restrained slabs are, in particular, often surprisingly thin. Therefore, in the many cases in which stiffness is important, it is necessary to have a thicker slab than is actually required for the bearing capacity. In order to decide this point,

n_a	2	2	1	2	1	0	1	0
n_b	2	1	2	0	1	2	0	1
α								
0·1	1·50	1·82	2·00	2·67	3·00	3·33	14·0	16·0
0·15	2·37	3·07	3·38	5·53	6·21	6·90	—	—
0·20	3·39	4·85	5·34	13·6	15·3	17·0	—	—
0·25	4·66	7·77	8·55	∞	∞	∞		
0·30	6·40	14·2	15·6	—	—	—		
0·35	9·10	45·5	50·0	—	—	—		
0·40	14·1	—	—	—	—	—		
0·45	29·0	—	—	—	—	—		
0·50	∞							

it is necessary to determine the maximum deflexion under the permissible load. As the modulus of elasticity of concrete may lie between 200 000 and 400 000 kg/cm^2 and as, moreover, the contribution from creep is rather uncertain, there is not much point in making a particularly accurate calculation. After all, it is not a case of deciding whether the deflexion is 5 mm or 6 mm, but whether it is 5 mm or 5 cm.

As the slab can be reckoned to be in the elastic stage under the working load, the deflexions can be calculated in accordance with the theory of elasticity. Unfortunately, this calculation is often rather inconvenient and, in practice, downright impossible in most cases. We therefore have to seek a simpler method.

We can now use the following property of elastic slabs. If we cut a narrow, straight strip out of a slab, we can determine its deflexions solely on the basis of the bending moments about axes perpendicular to the direction of the strip, whilst the bending and torsional moments acting on its sides are without influence provided we neglect the effect of Poisson's ratio. We get:

$$EI\,k_x = m_x$$

when the X-axis is located in the axis of the strip and $k_x = -\,\partial^2 u/\partial x^2$. It will be seen that the deflexions in the strip can be determined in the same way as for a beam. If the strip is now cut out in such a way that it is supported at both ends and contains the maximum moment in the slab, m_{max}, then the maximum deflexion will be

$$EIu = \alpha m_{max} l^2,$$

where l is the length of the strip and α is a constant which only depends upon the way in which the moment varies along the strip. If the moment is constant or varies linearly, then $\alpha \simeq 1/8$. If the moment curve is a parabola or a triangle with the apex at the centre, we get 1/9·6 or 1/12. It will be seen that the maximum deflexion can be expressed by m_{max}, and that the constant α lies between 1/8 and 1/12 when the moment curve lies between a rectangle and a triangle, which is normally the case.

Regarding m_{max}, we know that this must naturally be greater than the value of m corresponding to the permissible load because m is assumed to be constant along certain sections, whilst the theory of elasticity gives a variable moment. If, for example, the moment varies parabolically, its mean value will be $\frac{2}{3} m_{max}$. It can now be assumed that this mean value corresponds approximately to m, so $EIu \sim \frac{3}{2}\alpha m l^2$. The study of a large number of cases that can be calculated according to the theory of elasticity shows that we can reasonably substitute $\frac{3}{2}\alpha = \frac{1}{8}$.

In the case of statically restrained slabs, $u = u_o - u_i$, where u_o corresponds to simple support and u_i to the restraining moments only. According to the foregoing, $EIu_o \sim \frac{1}{8} m_o l^2$, where m_o is the moment for simple support. As $m = m_o - m_i$, the restraining moments must give m_i at the middle of the strip under consideration, so we can assume that $EIu_i \sim \frac{1}{8} m_i l^2$. We thus get $EIu = EI(u_o - u_i) \sim \frac{1}{8}(m_o - m_i) l^2 = \frac{1}{8} m l^2$.

As the restraining moments at the permissible load are usually greater than those assumed in the yield-line theory, which are used in the calculation of u_i, the value of u thereby determined will at any rate not be too low. Thus, by suitable choice of strip, we can estimate the maximum deflexion for both statically restrained and simply supported slabs by means of the following formula:

$$u = \frac{ml^2}{8EI} \qquad (1.17)$$

In the case of the rectangular slabs considered here, one yield line is parallel to the long sides, so m_{max} in the elastic stage must be reckoned to act in the same section. The strip must contain m_{max} and must therefore be orientated parallel to the short sides.

For simply supported rectangular slabs with a uniformly distributed load, the results of the theory of elasticity can, to a good approximation, be expressed as follows:

$$EIu = \frac{5}{384} \frac{pa^2b^2}{1 + \frac{a^2}{b^2} + \frac{b^2}{a^2}} \qquad (1.171)$$

The introduction of m from (1. 116″) yields

$$EIu = \frac{ma^2}{9{\cdot}6}\left[1 + \frac{\frac{b}{a} - \frac{a^2}{b^2}}{1 + \frac{a^2}{b^2} + \frac{b^2}{a^2}}\right] = \frac{ma^2}{K}$$

where K varies between 9·6 and 7·2, so that a suitable mean value is 8, which leads exactly to (1. 17).

In order to judge whether a calculated deflexion is permissible, two factors may be decisive, the absolute value of the deflexion and its order of magnitude in relation to the span of the slab. In a comparison with one-way slabs it must be remembered that these have their maximum deflexion over the entire section, whereas two-way slabs have it only at a single point. In cases in which the absolute magnitude of the deflexion is not decisive, we may perhaps content ourselves with keeping the mean deflexion under a certain limit, so the maximum deflexion can be 50% higher than that in one-way slabs. In structural engineering, for example, 1/500 of the span could be the upper limit for the maximum deflexion, when (1. 17) is used and E is reckoned as 200 000 kg/cm^2.

The question of which dimension of the slab is to be regarded as the span must be carefully considered in each individual case. In the case of the rectangular slabs dealt with in the foregoing, the shorter side should be taken as the span.

It may be useful to compare the stiffnesses of slabs having the same shape and method of support, but different size and load.

(a) Uniformly distributed load p. When l is one of the slab dimensions, we get, since $m = C_1pl^2$ and the thickness $h = c\sqrt{m} = c\sqrt{C_1p}\ l$:

$$u = \frac{3ml^2}{2Eh^3} = \frac{3}{2Ec^3\sqrt{C_1}} \frac{1}{\sqrt{p}};\ \frac{1}{u} = K_1\sqrt{p}$$

(b) Line load $\bar{p}$. As $m = C_2\bar{p}l$ and $h = c\sqrt{m}$, we obtain correspondingly

$$\frac{1}{u} = K_2\sqrt{\frac{\bar{p}}{l}}$$

(c) Point load P. As $m = C_3P$ and $h = c\sqrt{m}$, we obtain

$$\frac{1}{u} = K_3\frac{\sqrt{P}}{l}$$

As the stiffness increases with l/u, the latter can be taken as a measure of the stiffness. It will then be seen that the stiffness increases with the load in all cases. For the point loads, the stiffness decreases with the size of the slab.

The weakest slabs are therefore large slabs with a small but concentrated load.

Finally, we shall compare the stiffness in a one-way and a two-way, square slab, with the same span and uniformly distributed load. As both slabs are reckoned to be simply supported, we get:

For the one-way slab: the moment m_1, thickness $h_1 = c\sqrt{m_1}$, and the deflexion,

$$u_1 = \frac{5}{384}\frac{pl^4}{E \times \frac{1}{12}h_1^3}$$

For the two-way slab: $m = \frac{1}{3}m_1$, $h = c\sqrt{m}$, and the deflexion (in accordance with 1. 171),

$$u = \frac{5}{384}\frac{pl^4}{3E \times \frac{1}{12}h^3}$$

whence

$$\frac{u}{u_1} = \frac{1}{3}\frac{h_1^3}{h^3} = \sqrt{3}$$

This demonstrates clearly that two-way slabs have lower stiffness than one-way slabs. As technical advances have also entailed constantly increasing permissible stresses, it is clear that we must now also pay attention to the stiffness.

The deflexions calculated here correspond exactly to those forming the basis of an evaluation of the stiffness of steel and timber structures. They must not be confused with the permanent deflexions, which are caused by the shrinkage and creep of the concrete. A slab may well fulfil the stiffness requirements and still have large permanent deflexions, just as the reverse has, in fact, been observed in practice. The quality of workmanship plays a very large role as far as these permanent deflexions are concerned. On the basis of measurements on slabs constructed in practice, the permanent deflexion, with standard methods of construction, can be reckoned to be $y = \epsilon l^2/4h$, where $\epsilon = 0{\cdot}025 + 0{\cdot}0015 \times \sigma_b$ in per cent. Reckoning l in metres, σ_b in kg/cm^2, and h and y in cm, we get:

$$y = \frac{5 + 0{\cdot}3\,\sigma_b}{8h}l^2$$

Here, l is the span of the strip used for investigating the stiffness, and h is the thickness of the slab. Whether the concrete stress σ_b should correspond to dead load or to total load depends upon the actual conditions. These usually large deflexions are best avoided by careful construction, late stripping, and possibly by supporting the slab with a couple of props for as long as possible. The deleterious effect of the deflexions can be limited by erecting partition walls and similar as late as possible.

1.18 Orthotropic slabs

The calculation of orthotropic slabs can be converted, by means of the affinity theorem, to the calculation of isotropic slabs.

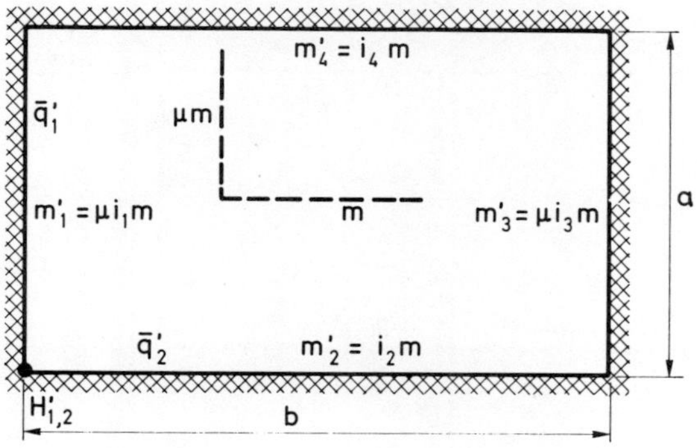

Fig. 1.18a

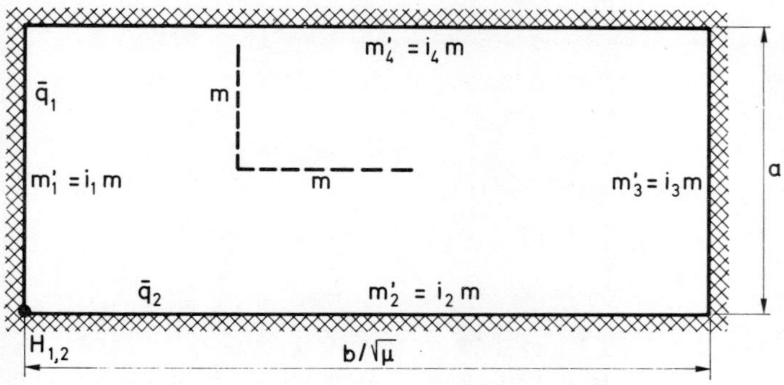

Fig. 1.18b

In the slab shown in Fig. 1.18a, where the sections parallel with b have the ultimate moment m, and the sections parallel with a, the ultimate moment μm, the moment m is the same as in an isotropic slab with the sides a and $b/\sqrt{\mu}$. In accordance with Fig. 1.18b, we then get

$$a_r = \frac{2a}{\sqrt{1+i_2} + \sqrt{1+i_4}}$$

$$b_r = \frac{2b}{\sqrt{\mu}(\sqrt{1+i_1} + \sqrt{1+i_3})} \qquad (1.181)$$

for the calculation of m by means of (1.116″).

For the reactions we get

$$\bar{q}_1 = \bar{q}'_1/\sqrt{\mu}, \bar{q}_2 = \bar{q}'_2$$

$$H_{1,2} = H'_{1,2}/\sqrt{\mu} \text{ etc.}$$

Substitution of $\bar{q}_1$, $\bar{q}_2$, $H_{1,2}$, etc., from (1.117) and (1.119) gives

$$\bar{q}'_1 = 4m\left(\frac{1}{a_r} + \frac{1}{b_r}\right)\sqrt{\mu}\sqrt{1+i_1}$$

$$\bar{q}'_2 = 4m\left(\frac{1}{a_r} + \frac{1}{b_r}\right)\sqrt{1+i_2}$$

$$H'_{1,2} = 2m\sqrt{1+i_1}\,\sqrt{1+i_2}\,\sqrt{\mu} \qquad (1.182)$$

1.19 Numerical example

In a warehouse with the dimensions 18·5 × 10·8 m, and p = 1000 kg/m², the floor shown in Fig. 1.19 consists of two-way slabs supported on beams. As the degree of restraint used is i = 1, we get the ratio 0·85 between the outer bays and intermediate bays when we use the same reduced span. Division into 3 and 4 bays gives respective spans of 10·8/(1 + 2 × 0·85) = 4·0 m and 18·5/(2 + 2 × 0·85) = 5·0 m for the intermediate bays. For the outer bays, we get 0·85 × 4·0 = 3·4 m and 0·85 × 5·0 = 4·25 m respectively. This division results in the same reduced spans for all slabs. Thus,

$$a_r = \frac{2 \times 4}{\sqrt{2} + \sqrt{2}} = \frac{4}{\sqrt{2}} = \frac{2 \times 3{\cdot}4}{1 + \sqrt{2}}; b_r = \frac{5}{\sqrt{2}} \qquad (1.111)$$

Assuming the dead load to be 220 kg/m², we get

$$m = \frac{1220 \times 4 \times 5}{2 \times 8(1 + 0{\cdot}8 + 1{\cdot}25)} = 500 \text{ kg} \qquad (1.116'')$$

The uniformly distributed reaction at a simply supported side is then

$$4 \times 500\left(\frac{\sqrt{2}}{4} + \frac{\sqrt{2}}{5}\right) = 1270 \text{ kg/m} \qquad (1.117)$$

and at a restrained side,

$$4 \times 500\left(\frac{\sqrt{2}}{4} + \frac{\sqrt{2}}{5}\right)\sqrt{2} = 1800 \text{ kg/m}$$

The negative corner reactions for corners with no restrained sides and with one and two restrained sides, respectively, are

$$2 \times 500 = 1000, 1000\sqrt{2} = 1410 \text{ and}$$
$$1000\sqrt{2}\sqrt{2} = 2000 \text{ kg} \qquad (1.119)$$

We thus get the loads on the beams shown in Fig. 1.19.

As a check, we have

$$1220 \times 10{\cdot}8 \times 18{\cdot}5 = 2(10{\cdot}8 + 18{\cdot}5)1270$$
$$+ (3 \times 10{\cdot}8 + 2 \times 18{\cdot}5)3600$$
$$- 4 \times 1000 - 10 \times 2820 - 6 \times 8000$$

With the stresses 75/1300 kg/cm², we get

$$h_n = 0{\cdot}260\sqrt{500} = 5{\cdot}81 \text{ cm}$$

$$f = 0{\cdot}348\sqrt{500} = 7{\cdot}8 \text{ cm}^2 \text{ per m}$$

This gives a slab thickness of 8 cm, which results in the following deflexion in the intermediate panels:

$$u = \frac{1/8 \times 500 \times 400^2}{200\,000 \times 1/12 \times 8^3} = 1{\cdot}17 \text{ cm}$$

which exceeds 400/500 = 0·8 cm. Therefore, the thickness of the slab should be increased in order to improve the stiffness. With 9 cm, we get

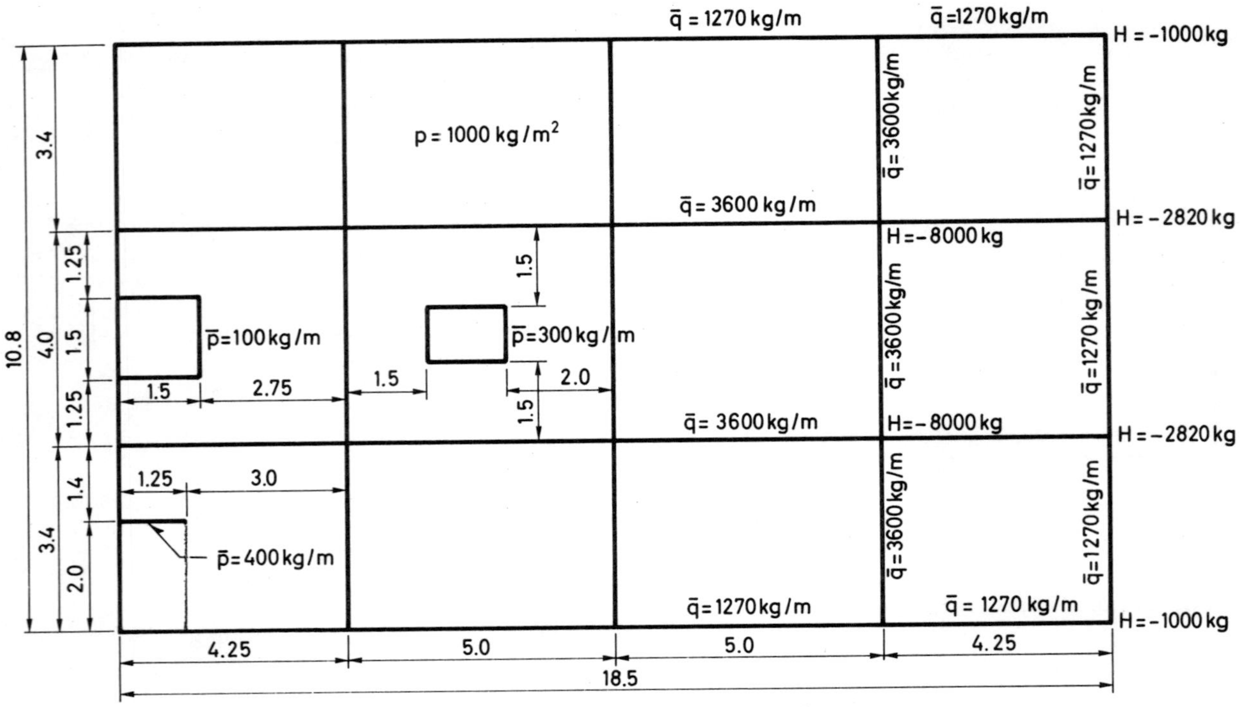

Fig. 1. 19

$u = 1{\cdot}17 \times (8/9)^3 = 0{\cdot}82$ cm, which is hardly sufficient, so a thickness of 10 cm should be selected. With 10 mm bars, we thereby get $h_n = 8{\cdot}0$ cm and $f = (7{\cdot}8/8{\cdot}0) \times 5{\cdot}81 = 5{\cdot}66$ cm² per m.

The extent of the reinforcement at the restrained edges can be found by means of (1. 163), which yields

$$500(4 - 8\alpha) \leq \alpha(1 - 1{\cdot}8\alpha + \tfrac{4}{3} \times 0{\cdot}8\alpha^2)240 \times 20$$

for slabs with four restraints. We find that $\alpha = 0{\cdot}3$, so the top reinforcement should extend $0{\cdot}3 \times 4 = 1{\cdot}2$ m into the slabs. The same value can be utilized in the outer panels, but the exact values can also be found.

For the outer panels with sides of 3·4 and 5 m, we get

$$500(3 - 8\alpha) = \alpha(1 - 1{\cdot}68\alpha + \tfrac{4}{3} \times 0{\cdot}68\alpha^2) \times 240 \times 3{\cdot}4 \times 5$$

i.e. $\alpha = 0{\cdot}22$. The distance is then $0{\cdot}22 \times 3{\cdot}4 = 0{\cdot}75$ m.

For the outer panels with sides of 4·0 and 4·25 m, we get

$$500(3 - 8\alpha) = \alpha(1 - 1{\cdot}94\alpha + \tfrac{4}{3} \times 0{\cdot}94\alpha^2) \times 240 \times 4 \times 4{\cdot}25$$

i.e. $\alpha = 0{\cdot}23$. The distance is then $0{\cdot}23 \times 4 = 0{\cdot}72$ m.

For the outer panels with sides of 3·4 and 4·0 m, we get

$$500(2 - 8\alpha) = \alpha(1 - 1{\cdot}85\alpha + \tfrac{4}{3} \times 0{\cdot}85\alpha^2) \times 240 \times 3{\cdot}4 \times 4{\cdot}0$$

i.e. $\alpha = 0{\cdot}16$. The distance is then $0{\cdot}16 \times 3{\cdot}4 = 0{\cdot}55$ m.

According to (1. 166), we need $\alpha = 0{\cdot}17$ in both directions, i.e. 0·57 and 0·71 m, respectively. If we take the mean value 0·64 in both directions, we get the following value of m in the middle panel of the slab:

$$m = \frac{1220 \times 2{\cdot}76 \times 3{\cdot}61}{8\left(1 + \frac{2{\cdot}76}{3{\cdot}61} + \frac{3{\cdot}61}{2{\cdot}76}\right)} = 495 < 500 \text{ kg}$$

In one of the intermediate panels, there is a hole as shown in Fig. 1. 19. The line load along the edge of the hole is $\bar{p} = 300$ kg/m. As we are only going to strengthen the bottom reinforcement, the restraining moment must still be reckoned as 500 kg. With $mi_1 = mi_2 = mi_3 = mi_4 = 500$ kg, $a = 4$, $b = 5$, $c_1 = c_2 = c_4 = 1{\cdot}5$, and $c_3 = 2{\cdot}0$, equation (1. 151) gives

$$500\left(\frac{4}{1{\cdot}5} + \frac{4}{2{\cdot}0} + \frac{5}{1{\cdot}5} + \frac{5}{1{\cdot}5}\right) + m \times 3{\cdot}5 \times \frac{2}{1{\cdot}5} + m \times 3{\cdot}0\left(\frac{1}{1{\cdot}5} + \frac{1}{2{\cdot}0}\right)$$

$$= \tfrac{1}{2} \times 1240(4 \times 5 - 1 \times 1{\cdot}5) + 300 \times 5 - \tfrac{1}{6} \times 1240 \times 3{\cdot}5 \times 3{\cdot}0$$

This results in $m = 641$ kg, $h_n = 0{\cdot}260\sqrt{641} = 6{\cdot}59$ cm and $f = 0{\cdot}348\sqrt{641} = 8{\cdot}80$ cm² per m.

With $h_n = 8{\cdot}0$ cm, we get $f = 8{\cdot}80 \times (6{\cdot}59/8{\cdot}0) = 7{\cdot}25$ cm² per m.

In one of the outer panels, which measures 4·0 × 4·25 m, there is a lift well measuring 1·5 × 1·5 m, as shown, and with $\bar{p} = 100$ kg/m on the edge. With

$i_2 = 0$, $mi_1 = mi_3 = mi_4 = 500$ kg, a = 4·25, b= 4·00, $c_1 = c_3 = 1{\cdot}25$, c = 1·5, and $c_4 = 2{\cdot}75$, equation (1. 157) yields

$$4{\cdot}25(m + 500)\,\frac{2}{1{\cdot}25} + m \times 2{\cdot}5\left(\frac{1}{1{\cdot}5} + \frac{1}{2{\cdot}75}\right)$$

$$+ 500 \times 2{\cdot}5 \times \frac{1}{2{\cdot}75}$$

$$= \tfrac{1}{6} \times 1240 \times 4{\cdot}25 \times 9{\cdot}5 - \tfrac{1}{2} \times 1240 \times 1{\cdot}5^2$$

$$+ 100 \times 3$$

We get m = 500 kg, so no alteration to the reinforcement is called for here. Finally, in the corner panel, there is to be a stairway opening measuring 2·0 × 1·25 m, with $\bar{p} = 400$ kg/m on the short edge. It is obviously formula (1. 156) that should be used. With a = 3·4, b = 4·25, $i_1 = i_2 = 0$, $mi_3 = mi_4 = 500$ kg, c = 2·0, d = 1·25, we then get

$$m \times \frac{1{\cdot}4}{1{\cdot}25} + m \times \frac{3{\cdot}0}{2{\cdot}0} + (m + 500)\left(2 + \frac{4{\cdot}25}{1{\cdot}4}\right)$$

$$= \frac{1240}{12} \times 3{\cdot}4(6 \times 4{\cdot}25 - 3{\cdot}4)$$

$$- \frac{1240}{6} \times 1{\cdot}25 \times 7{\cdot}4 + \frac{1}{2} \times 500$$

from which we get m = 467 kg, so the reinforcement can be reduced to (5·66/500) × 467 = 5·3 cm^2 per m.

1. 2 Supported along three sides

1. 21 Uniformly distributed load and line load

We will at once derive equations for the uniformly distributed load p and the line load $\bar{p}_a$ parallel with side a and $\bar{p}_b$ on the free edge. The yield pattern shown in Fig. 1. 211 gives the following moment equations about the edges for the parts of the slab shown, the nodal forces at the edge being $m(h_1/a)$ and $m(h_3/a)$:

For I:

$$m(1 + i_1)a - m\,\frac{h_1}{a}\,h_1 = \frac{1}{6}\,pah_1{}^2 + \frac{1}{2}\,\bar{p}_b h_1{}^2$$

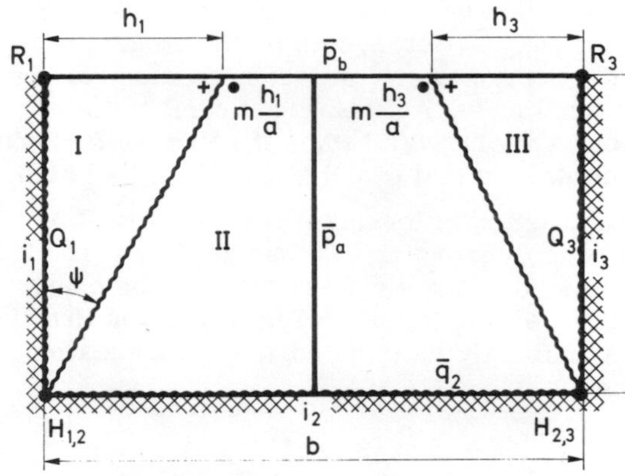

Fig. 1. 211

or

$$m\left(1 + i_1 - \frac{h_1{}^2}{a^2}\right) = \frac{1}{6}\,ph_1{}^2\left(1 + 3\,\frac{\bar{p}_b}{pa}\right)$$

and correspondingly for III:

$$m\left(1 + i_3 - \frac{h_3{}^2}{a^2}\right) = \frac{1}{6}\,ph_3{}^2\left(1 + 3\,\frac{\bar{p}_b}{pa}\right)$$

If we introduce $h_1 = h\sqrt{1 + i_1}$ and $h_3 = h\sqrt{1 + i_3}$, both equations yield

$$m\left(1 - \frac{h^2}{a^2}\right) = \frac{1}{6}\,ph^2\left(1 + 3\,\frac{\bar{p}_b}{pa}\right) = \frac{1}{6}\,ph^2(1 + 3\beta) \qquad (1)$$

For II we get the moment equation

$$i_2 mb + m(h_1 + h_3) + m\,\frac{h_1}{a}\,a + m\,\frac{h_3}{a}\,a$$

$$= \tfrac{1}{2}pba^2 - \tfrac{1}{3}p(h_1 + h_3)a^2 + \tfrac{1}{2}\bar{p}_a a^2$$

$$+ \bar{p}_b(b - h_1 - h_3)a$$

Introducing here h and $b_r = 2b/(\sqrt{1 + i_1} + \sqrt{1 + i_3})$, we get

$$m\left(\frac{i_2}{4} + \frac{h}{b_r}\right)$$

$$= \frac{1}{24}\,pa^2\left[3 - 4\,\frac{h}{b_r} + 3\alpha + 6\beta\left(1 - 2\,\frac{h}{b_r}\right)\right] \qquad (2)$$

where α and β have the same meaning as in 1. 13.

Multiplying (1) by a/h and (2) by b_r/a, we get, by addition,

$$m\left(\frac{i_2}{4}\,\frac{b_r}{a} + \frac{a}{h}\right) = \frac{1}{8}\,pab_r(1 + \alpha + 2\beta) \qquad (3)$$

From (1) and (3) we get a second-order equation for a/h, so that

$$\frac{a}{h} = K + \sqrt{K^2 + \frac{1}{2}\,i_2\,\frac{b_r}{a}\,K + 1} \qquad (4)$$

$$K = \frac{2}{3}\,\frac{1 + 3\beta}{1 + \alpha + 2\beta}\,\frac{a}{b_r}$$

after which we find m from (3).

It will be seen that the calculation of slabs restrained along the sides a is thus converted to the calculation of slabs that are simply supported along the sides a. As earlier, we do this by introducing the reduced span b_r. In the same way, the calculation of slabs subjected to a line load is converted to the calculation of slabs without a line load by means of the following transformation:

$$p' = p(1 + \alpha + 2\beta);\; i_2{}' = i_2\,\frac{1 + 3\beta}{1 + \alpha + 2\beta}$$

$$a' = a\sqrt{\frac{1 + 3\beta}{1 + \alpha + 2\beta}};\; b' = b_r\sqrt{\frac{1 + \alpha + 2\beta}{1 + 3\beta}} \qquad (1.\,211)$$

which is closely related to the transformations in 1.13.

In particular, $\alpha = \beta$ again gives the same slab, but with $p' = p(1 + 3\alpha)$.

For the slab without a line load, comparative calculations show that the moment is closely approximated by the following formula, which is somewhat on the safe side:

$$m = \frac{p'a'b'}{8 + 6\frac{a'}{b'} + 2\left(\frac{a'}{b'} + \frac{b'}{a'}\right)i_2'} \quad \text{for } a' \leq b' \qquad (1.212)$$

When $a' \leq b'$, the value of m given by this formula is at most 7% on the safe side. For higher values of a', the formula is no longer of interest because the yield pattern then becomes as shown in Fig. 1.212. Before dealing with this pattern, we should note an important difference between the slabs with $i_2 = 0$ and the slabs with $i_2 \neq 0$.

In the first case, when b_r is large in relation to a, and a/b_r and α both approach zero, the limiting value of m is

$$m \rightarrow \tfrac{1}{8}pab_r + \tfrac{1}{4}\bar{p}_b b_r \text{ for } i_2 = 0$$

which approaches ∞ when $b_r \rightarrow \infty$. In the other case, with $m_2' = i_2 m$, we get

$$m_2' \rightarrow \tfrac{1}{2}pa^2 + \bar{p}_b a \text{ for } i_2 \neq 0$$

which is also found directly as the moment in a cantilever slab.

This moment is finite and is therefore of a completely different order of magnitude than the moment found above for $i_2 = 0$.

As mentioned, higher values of a will result in the yield pattern shown in Fig. 1.212. The transition from the first yield pattern to the latter does not occur when the two yield lines meet at the edge, i.e. when $h_1 + h_3 = b$, but before this, i.e. when the value of m determined by means of Fig. 1.212 is equal to that determined by means of Fig. 1.211.

If the transition did occur in the former manner, we would get nodal forces at the edge in the transition pattern if this were assumed to have arisen from Fig. 1.211, but none if it were assumed to have arisen from Fig. 1.212. This discrepancy shows that the transition cannot occur geometrically ($h_1 + h_3 = b$), but must occur statically (maximum m is decisive).

The calculation of the yield pattern in Fig. 1.212 can immediately be related to the calculation of the rectangular slab with sides 2a and b, supported on all four sides, with the degree of restraint $i_4 = i_2$, and loading p, $\bar{p}_a$ and $2\bar{p}_b$, whereby α and β remain unaltered. If, namely, we imagine this slab to be cut along the symmetry line through the middle of the 2a's, we get the given slab with free edge and a corresponding, symmetrical slab. The only difference between these two slabs and the slab that is twice as

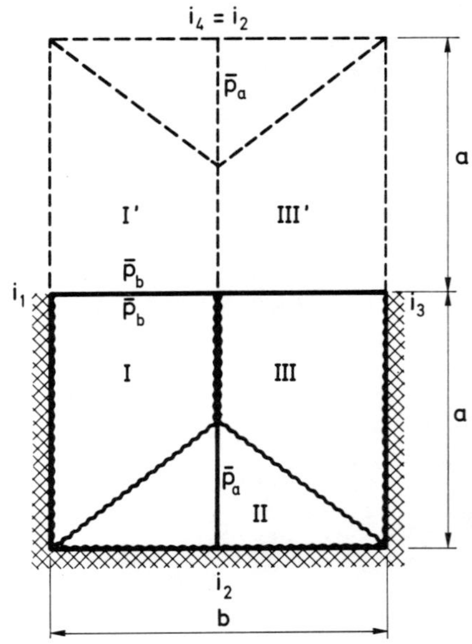

Fig. 1.212

large is that there is no bending moment in the section in question in the former, whereas there is a certain moment in the latter. However, this difference has no effect upon the equilibrium equations for I + I' and III + III', as these are moment equations about the edges 2a, which are perpendicular to the section in question, so its bending moment has no component in the direction of the side 2a. Because of the symmetry there are no transverse forces or torsional moments in the section at the plane of symmetry. Using formulae (1.13) and noting that the side parallel to the middle yield line should be denoted b″ (Fig. 1.111) and that $a_r = a/\sqrt{1 + i_2}$, we get

$$p'' = p(1 + \beta + 2\alpha) = p'; \; a'' = b_r$$

$$b'' = \frac{2a}{\sqrt{1 + i_2}} \sqrt{\frac{1 + \beta + 2\alpha}{1 + 3\alpha}} \qquad (1.213)$$

after which m is calculated from (1.116″), with p″, a″ and b″. As both this formula and formula (1.212) are approximate, they cannot be utilized for determining the transitional case, but the value given for (1.212) can always be used. As m does not vary much in different yield patterns, both formulae give practically the same value of m for cases close to the transitional case. As mentioned earlier, these calculations are only valid when the line loads do not predominate, so that α and β are of limited size.

Other approximate formulae can be derived by methods corresponding to those used in 1.11. If it is assumed that the yield lines follow the corner bisectors corresponding to Fig. 1.112 and that the load is uniformly distributed, the work equation becomes

$$m(1 + i_1)a + 2ma + mi_2b + m(1 + i_3)a = \tfrac{1}{2}\,pab - \tfrac{1}{3}\,pa^2$$

or

$$m = \frac{pab}{8} - \frac{pa^2}{12} - \frac{m}{4}(i_1 + i_2\frac{b}{a} + i_3) = m_o - m_i$$

$$m_o = \frac{pab}{8}\left(1 - \frac{2}{3}\frac{a}{b}\right); m_i = \frac{a(m_1' + m_3') + bm_2'}{4a}$$

However, the approximation is not as good here as in 1.11 because the error may be as much as 10%, and on the unsafe side. When $b < 2a$, we get, corresponding to (1.116′), with $i_2 = i_4$ and the sides 2a and b, of which 2a is the greater,

$$m_o = \frac{pb^2}{24}\frac{6a - b}{2a + b}; m_i = \frac{a(m_1' + m_3') + bm_2'}{2a + b}$$

By utilizing our knowledge of the distribution of the reactions, we can, as in 1.11, derive a safe approximate formula. By means of the transformation (1.211) or (1.213), we reduce the problem to that of a slab a′ b′, which is simply supported along a′ and subjected to the uniformly distributed load p′. According to YLT, §11.3, page 139, the reaction Q_b on b′ is uniformly distributed. On a′, according to §11.4, page 140, we get a concentrated reaction $R_o = 2m$ at the corner at the free edge; the remaining reaction Q_a is distributed as shown in Fig. 1.213. Let us imagine the slab to be supported on simple beams with the moments M_a and M_b. A section parallel with a′ through the middle of the slab and the beam at b′ then immediately gives

$$ma' + M_b = \tfrac{1}{8}p'a'(b')^2 = ma' + \tfrac{1}{8}Q_b b' \qquad (1)$$

The moment equation for side b′ gives

$$2R_o a' + 2Q_a z + i_2' b' m = \tfrac{1}{2}p'(a')^2 b' \qquad (2)$$

and, further,

$$p'a'b' + 2H = 2Q_a + Q_b + 2R_o \qquad (3)$$

Q_a and Q_b can be found from (1) and (2), and $R_o = 2m$. The corner forces H are $m(1 + i_2')\cot\varphi + m\tan\varphi$, the minimum value of which is $2m\sqrt{1 + i_2'}$. Inserting these in (3), we get

$$m \geq \frac{p'a'b'}{8 + 16\frac{z}{b'} + 2i_2'\frac{b'}{a'} + 8\frac{z}{a'}(\sqrt{1 + i_2'} - 1)} \qquad (4)$$

It will be seen from the distribution shown for Q_a that z must be slightly greater than $\tfrac{1}{3}a'$. If we take $z = \tfrac{3}{8}a'$, we get from (4) a formula that is identical with (1.212), apart from the last term in the denominator.

On the same basis, the reactions for the given slab can also be determined, although only for $\bar{p}_a = 0$ as it is only in this case that Q_b is uniformly distributed. We find, for example (Fig. 1.211), $H_{1,2} = m(1 + i_1)\tan\varphi + m(1 + i_2)\cot\varphi$, which has minimum $2m\sqrt{1 + i_1}\sqrt{1 + i_2}$. According to YLT, page 141, we find, for instance, that $R_1 = 2m\sqrt{1 + i_1}$. The slab

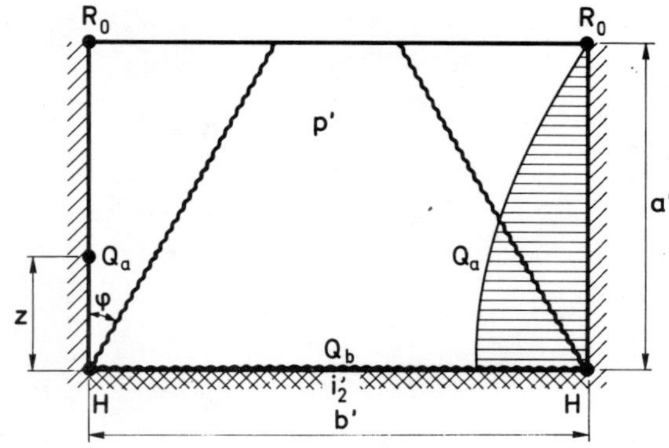

Fig. 1.213

and supporting beam are considered as a statically restrained beam subjected to a uniform load $pa + \bar{p}_b - \bar{q}_2$, whereby $ma = \tfrac{1}{8}(pa + \bar{p}_b - \bar{q}_2)b_r^2$. From this we find $\bar{q}_2$. On the same basis, formulae (0.12), for example, give $R_1 + Q_1 - H_{1,2} = \tfrac{1}{2}(pa + \bar{p}_b - \bar{q}_2)b_r\sqrt{1 + i_1}$, which determines Q_1. By this means we find that

$$\bar{q}_2 = pa + \bar{p}_b - \frac{8ma}{b_r^2}; H_{1,2} = 2m\sqrt{1 + i_1}\sqrt{1 + i_2}$$

$$Q_1 = 2m\left(\sqrt{1 + i_2} - 1 + 2\frac{a}{b_r}\right)\sqrt{1 + i_1};$$

$$R_1 = 2m\sqrt{1 + i_1} \qquad (1.214)$$

and corresponding expressions for R_3 and Q_3. The Q's are distributed in such a way that they give the moment $\tfrac{1}{8}Qa$ in a simple beam (YLT, page 142). For the yield pattern in Fig. 1.212, it is only the side of length 2a that is subject to the same reaction as that of the equivalent rectangular slab because the missing bending moment in the section at the plane of symmetry affects the calculations for determining the distribution of the reactions. As neither of the two yield patterns is specifically assumed in the considerations leading to the equations in (1.214), these equations apply to both cases.

In the foregoing, the special conditions applying at the corners (YLT, §8) are neglected. The calculated corner forces thus assume top reinforcement at the corner. In the case of a simply supported corner, this corresponds to $m' \sim m$ (YLT, page 91, formula (55), with $\omega = 90°$ and $u \sim 45°$), which is easily achieved.

At the restrained corner, a much higher degree of reinforcement is required, which is hardly practicable, so the assumed yield patterns are not completely correct in this case. The yield lines cannot pass through the corners, but corner levers are formed, which are of much greater importance in this type of slab than in the type in 1.1. This is because the vertices of the corner levers fall outside the free edge (Fig. 1.214a). The slabs that could previously be considered as one half of slabs with four supported sides can no longer be regarded in this manner. With four sides supported, we get corner levers of the type shown in Fig. 1.214 b, while in the corresponding slab with three sides

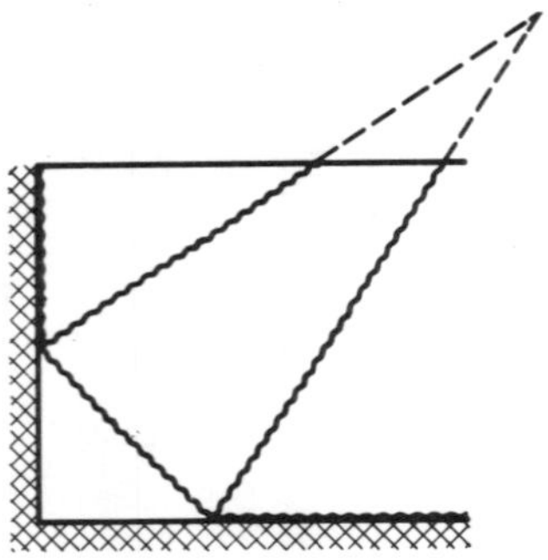

Fig. 1. 214a

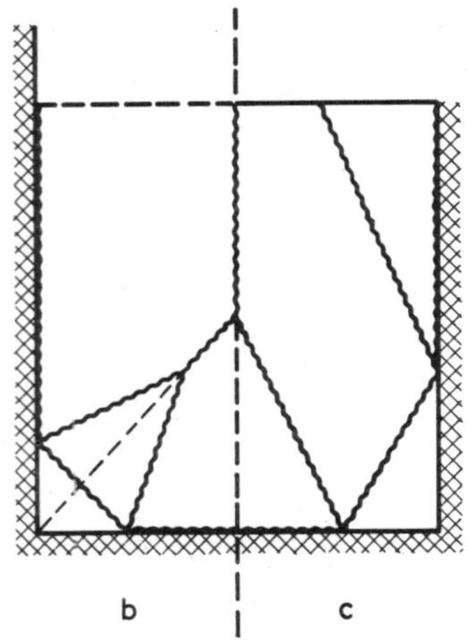

Fig. 1. 214b and c

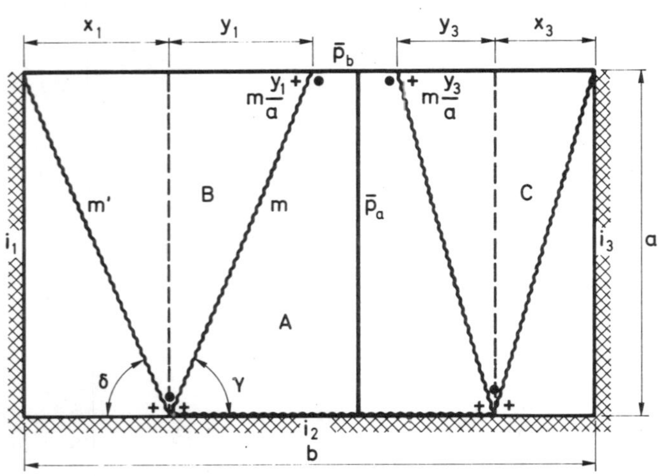

Fig. 1. 215

supported, we get corner levers of the type shown in Fig. 1. 214c on account of the free edge. However, it is very difficult to calculate these yield patterns, so we select instead the pattern shown in Fig. 1. 215.

At the free edge we find the nodal forces $m(y_1/a)$ and $m\ (y_3/a)$. At the opposite edge, in accordance with Fig. 46 in YLT, the nodal force for slab part B is found to be

$$m \cot\gamma + m_b' \cot\gamma + m_v'$$

where, in accordance with (28) and (29), page 52 in YLT,

$$m_b' = m_2' \cos^2\gamma + m_1' \sin^2\gamma$$

$$m_v' = (m_2' - m_1') \sin\gamma \cos\gamma$$

since the slab is orthotropically reinforced at the top, corresponding to m_1' and m_2'. With these values, the nodal force at B becomes

$$(m + m_2') \cot\gamma = m(1 + i_2)\frac{y_1}{a}$$

This is also obtained by noting that the edge nodal forces Q_R at the same point are $(m_2' - m_2') \cot\delta =$ 0 and $(m + m_2') \cot\gamma$ (YLT, page 51) and then applying (26), page 48 in YLT, which states that the sum of the nodal forces at the point is zero.

Correspondingly, for the nodal force at slab part C, we get $m(1 + i_2)(y_3/a)$. With these nodal forces, moment about the free edge for slab part B gives

$$my_1 + m(1 + i_2)\frac{y_1}{a}a - m_2'x_1$$

$$= m[(2 + i_2)\ y_1 - i_2x_1] = \frac{1}{6}pa^2(x_1 + y_1) \qquad \text{(B1)}$$

For the same slab part, moment about side a gives

$$ma + m_1'a + m(1 + i_2)\frac{y_1}{a}x_1 - m\frac{y_1}{a}(x_1 + y_1)$$

$$= \frac{1}{2}pa(x_1 + y_1)\frac{2x_1 + y_1}{3} + \frac{1}{2}\bar{p}_b(x_1 + y_1)^2$$

or

$$m[(1 + i_1)a^2 + y_1(i_2x_1 - y_1)]$$

$$= \frac{1}{6}pa^2(x_1 + y_1)(2x_1 + y_1) + \frac{1}{2}\bar{p}_b a(x_1 + y_1)^2 \qquad \text{(B2)}$$

For slab part C we get corresponding expressions with the suffix 3 instead of suffix 1. If we now substitute $x_1 = x\sqrt{1+i_1}$; $y_1 = y\sqrt{1+i_1}$; $x_3 = x\sqrt{1+i_3}$ and $y_3 = y\sqrt{1+i_3}$, we get the same equations for the two parts B and C of the slab, i.e. the calculation of the slab restrained at the sides a is converted to the calculation of the slab simply supported along a. We get

$$m[(2+i_2)y - i_2x] = \frac{1}{6}pa^2(x+y) \qquad \text{(b1)}$$

$$m[a^2 + y(i_2x - y)] = \frac{1}{6}pa^2(x+y)[(2+3\beta)x + (1+3\beta)y] \qquad \text{(b2)}$$

where, as previously, β is $\bar{p}_b/pa$.

For slab part A, we get the moment equation about side 2

$$m(y_1+y_3) + m\frac{y_1}{a}a + m\frac{y_3}{a}a + m_2'(b - x_1 - x_3)$$
$$= m[2(y_1+y_3) + i_2(b - x_1 - x_3)]$$
$$= \frac{1}{2}pa^2(b - x_1 - x_3) - \frac{1}{3}pa^2(y_1+y_3)$$
$$+ \bar{p}_b a(b - x_1 - x_3 - y_1 - y_3) + \frac{1}{2}\bar{p}_a a^2 \qquad \text{(A)}$$

The introduction of x, y and b_r leads to

$$m[4y + i_2(b_r - 2x)]$$
$$= \frac{1}{6}pa^2[3b_r(1+\alpha+2\beta) - 6(1+2\beta)x - 4(1+3\beta)y] \qquad \text{(a)}$$

where, as previously, α is $\bar{p}_a/pb$.

By eliminating x and y from these equations, we get an equation for determination of m. From (b1) and (a), which are linear in x and y, we find, by introducing $k = pa^2/6m$,

$$x = \frac{2 + i_2 - k}{i_2 + k}\, y$$

$$y = \frac{1}{2}b_r(k + i_2) \times \frac{3(1+\alpha+2\beta)k - i_2}{2k(4+3i_2) + 12\beta k(1+i_2) - k^2 - i_2^2} \qquad (1.215)$$

When these expressions are introduced in (b2), we get, after rearrangement, the following quadratic equation for determination of k and thus of m:

$$k^2\left[9(1+i_2)(1+\alpha+2\beta)^2 + 4\frac{a^2}{b_r^2}\right]$$
$$- 2k[3(1+\alpha+2\beta)(1+i_2)i_2 + 4\frac{a^2}{b_r^2}(4 + 3i_2 + 6\beta(1+i_2))]$$
$$+ i_2^2\left(1 + i_2 + 4\frac{a^2}{b_r^2}\right) = 0 \qquad (1.216)$$

When solving this equation by the usual formula, the square root must be taken as positive.

In the special case of the slab simply supported along b, we get:

$$m = \frac{3p}{32(2+3\beta)}\left[\frac{4}{9}a^2 + (1+\alpha+2\beta)^2 b_r^2\right]$$

$$x = \frac{3}{4}\,\frac{1+\alpha+2\beta}{2+3\beta}\,a - \frac{1+2\beta}{2+3\beta}\,\frac{1}{1+\alpha+2\beta}\,\frac{a^2}{b_r}$$

$$y = \frac{2}{3}\,\frac{1}{1+\alpha+2\beta}\,\frac{a^2}{b_r}$$

(1.217)

When b and thereby b_r are large in relation to a, m will approach

$$\frac{3p}{32(2+3\beta)}(1+\alpha+2\beta)^2 b_r^2$$

i.e. a moment of quite a different order of magnitude from that obtained in the case of the restrained slab (1.212).

When $i_2 > 0$, we cannot, as earlier, specify a simple transformation that will relate the general case to the special case $\alpha = 0, \beta = 0$. In the latter case, which is the most frequent, comparative calculations show that m is approximated by the following formula, which is only valid for slabs restrained along b:

$$m = \frac{pab_r}{3 + 12\frac{a}{b_r} + 2\left(1 + \frac{b_r}{a}\right)i_2} \qquad (1.218)$$

Formulae (1.216) to (1.218) are only valid when b_r exceeds a, and only in such cases do the corner levers have a major influence. For slabs where a exceeds b_r, it is sufficiently accurate to reckon on an increment of 10% in the value of m determined earlier in the case of anchored slabs (1.213) and (1.116″), as we saw (Fig. 1.214b and c) that the influence of the corner levers here was slightly greater than the 8% for slabs with four supported sides.

As already mentioned on page 11, more dangerous yield patterns arise for the type of slabs treated there, and as Fig. 1.216 shows, such a pattern would be of precisely the type dealt with here (Fig. 1.215), with a = c and b = l. Without top reinforcement, we then get from (1.217), with $\alpha = 0$,

$$m = \frac{3p}{32(2+3\beta)}\left[(1+2\beta)l^2 + \frac{4}{9}c^2\right];\ \beta = \frac{p}{pc} \qquad (1.219)$$

which must be used if it yields a higher value of m than (1.152) to (1.154). In the case of top reinforce-

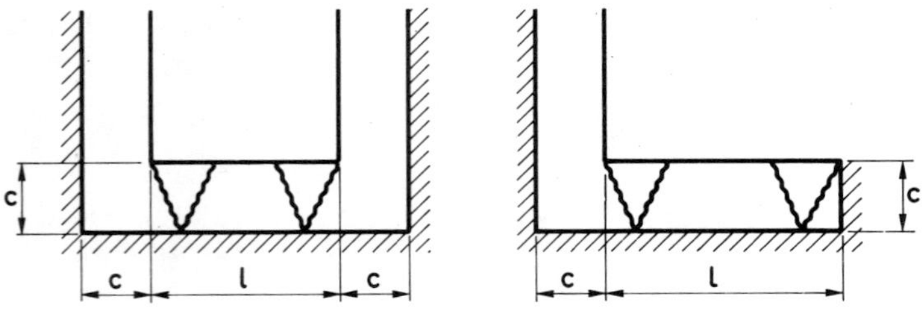

Fig. 1.216

ment, l_r must be used instead of l. By comparison with (1.152), we thus obtain

$$\frac{p}{16}[2c(a + b) - 4c^2] - \frac{pc^2}{12} < \frac{3p}{64}[(b - 2c)^2 + \frac{4}{9}c^2]$$

$$b \geq a$$

as conditions for the validity of (1.219). This gives

$$\frac{b}{c} > \frac{4}{3}\left[\frac{a}{b} + 2{\cdot}5 + \sqrt{\frac{a^2}{b^2} + 5\frac{a}{b} - 0{\cdot}25}\right]$$

which varies from 6·1 to 7·9 when a/b varies from 0·5 to 1.

The other two equations can be dealt with in a similar manner. In the case of (1.154), we get, correspondingly, $b/c > 3{\cdot}0$–$4{\cdot}7$, whilst in the case of (1.153), we get $b/c > 5{\cdot}0$ for both $a/b = 0{\cdot}5$ and $a/b = 1$. When $a = 2b$, we get $b/c > 2$, i.e. (1.219) must always be used in this case.

When the slabs discussed in this section are provided with a parapet at the free edge and this is to be calculated for a horizontal load v per m, the free edge gets the moment $m_R = vh$, where h is the height of the parapet. There is no difficulty in including this in the equations derived in connexion with Fig. 1.211, and thereby finding its influence on m. As m_R is usually small in relation to m, we can use the law of superposition without being too conservative, so in this case, the slab is calculated for $m + m_R$.

1.22 *Triangular loading*

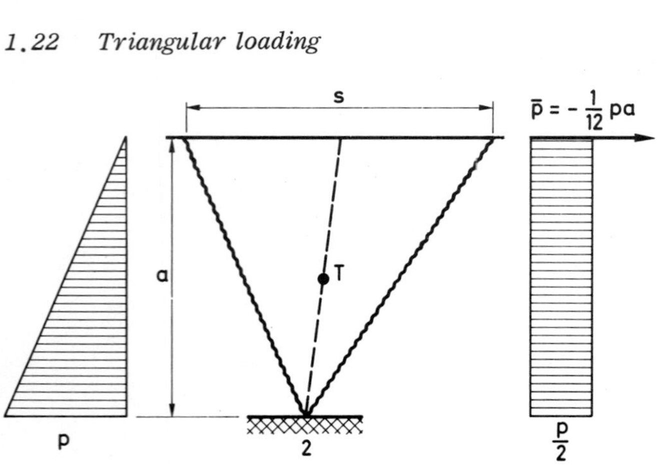

Fig. 1.221

Here we will only consider the triangular loading that arises in practice, which is zero along the free edge and p at the opposite edge.

When the yield patterns are as shown in Figs. 1.211 and 1.215, we get triangular slab parts with the baseline s on the free edge and height a, and a trapezoidal part. For the former slab parts we formulate moment equations about two axes, and for the latter slab part only about side 2. Fig. 1.221 shows one of the triangles mentioned, with a load equal to $\frac{1}{6}$ pas, acting at the mid-point T of the median. On the right of the same Figure is shown a uniformly distributed load $\frac{1}{2}$p and a line load $-\frac{1}{12}$pa acting along the free edge. A simple calculation shows that this combined load results in exactly the same total load acting at exactly the same point T as the triangular load (total load is $\frac{1}{2}p \times \frac{1}{2}as - \frac{1}{12}pa \times s = \frac{1}{6}pas$; the moment equation about 2 is $\frac{1}{4}pas \times \frac{2}{3}a - \frac{1}{12}pas \times a = \frac{1}{12}pa^2s$, i.e. T is located a/2 from 2 and must clearly also lie on the median). It is therefore immaterial which of the two loads we use for all equilibrium conditions for this part of the slab. The trapezoidal part of the slab can be divided into a rectangular and two (negative) triangles, as above. In the case of the triangles, the triangular load can be replaced by the combined load. In the case of the rectangle, the triangular load and the combined load have the same moment about side 2, namely, $\frac{1}{6}pa^2 = \frac{1}{4}pa^2 - \frac{1}{12}pa^2$ per unit of length. As we only use this moment equation for this particular part of the slab, it follows that the two loads result in the same moment equations for all parts of the slab. We have thus proved that, for the yield patterns in question, the triangular load can be replaced by a uniformly distributed load $\frac{1}{2}$ p and the edge load $\bar{p}_b = -\frac{1}{12}pa$; i.e. $\beta = -\frac{1}{6}$.

Then, for an anchored slab with $i_2 = 0$, we obtain from (1.212) and the transformation (1.211):

$$m = \frac{\frac{1}{2}pab_r}{12 + 6{\cdot}75\frac{a}{b_r}} \qquad (1.221)$$

When the slab is not anchored or when $i_2 > 0$, equation (1.216) must be used. This gives

$$4k^2\left(1 + i_2 + \frac{a^2}{b_r^2}\right) - 4k\left[i_2(1 + i_2) + 2(3 + 2i_2)\frac{a^2}{b_r^2}\right] + i_2^2\left(1 + i_2 + 4\frac{a^2}{b_r^2}\right) = 0 \qquad (1.222)$$

with $k = pa^2/12m$. In the special case in which $i_2 = 0$, we find

$$m = \frac{p}{72}(a^2 + b_r^2) \tag{1.223}$$

When $i_2 > 0$, comparative calculations show that we have, to a good approximation,

$$m = \frac{\frac{1}{2} pab_r}{6 + 16\frac{a}{b_r} + \left(2 + 3\frac{b_r}{a}\right)i_2} \tag{1.224}$$

For the yield pattern corresponding to Fig. 1.222, there is no longer any advantage in considering the slab as half of a slab supported on four sides, because the load is no longer uniformly distributed. We will therefore formulate moment equations for the slab parts shown. Now the load on a strip at a distance x from the free edge is p(x/a), and we get

$$\text{I:} \quad m(1 + i_1)a = \frac{1}{2}\int_0^{a-h} p\frac{x}{a} h_1^2 dx + \frac{1}{2}\int_{a-h}^{a} p\frac{x}{a}\left(h_1 \frac{a-x}{h}\right)^2 dx$$

and

$$\text{II:} \quad m(1 + i_2)b = \int_{a-h}^{a} p\frac{x}{a}\frac{b}{h}(h - a + x)(a - x)\, dx$$

and a corresponding moment equation for III. We then find

$$m(1 + i_1) = \frac{p}{24} h_1^2\left(6 - 8\frac{h}{a} + 3\frac{h^2}{a^2}\right), h_1 + h_3 = b$$

$$m(1 + i_3) = \frac{p}{24} h_3^2\left(6 - 8\frac{h}{a} + 3\frac{h^2}{a^2}\right) \tag{1.225}$$

$$m(1 + i_2) = \frac{pa^2}{12}\left(2 - \frac{h}{a}\right)\frac{h^2}{a^2}$$

These equations show that the calculation of the restrained slab can be converted to the calculation of the simply supported slab by inserting $h_1 = \frac{1}{2} b_r \sqrt{1 + i_1}$ and $h_3 = \frac{1}{2} b_r\sqrt{1 + i_3}$, where b_r is the usual reduced span and $a_r = a/\sqrt{1 + i_2}$. Thence,

$$m = \frac{p}{96} b_r^2\left(6 - 8\frac{h}{a} + 3\frac{h^2}{a^2}\right)$$

$$m = \frac{p}{12} a_r^2\left(2 - \frac{h}{a}\right)\frac{h^2}{a^2} \tag{1.226}$$

which is precisely the form taken by (1.225) when all $i = 0$ and $h_1 = h_3 = b/2$.

By division we get

$$8\left(\frac{a_r}{b_r}\right)^2 = \frac{6 - 8\frac{h}{a} + 3\frac{h^2}{a^2}}{\left(2 - \frac{h}{a}\right)\frac{h^2}{a^2}} \tag{1.227}$$

which, together with one of the equations (1.226), gives a parametric representation of the relation between m and a_r/b_r, with h/a as parameter. Comparative calculations then show that the following is an excellent approximation:

$$m = \frac{\frac{1}{2} pa_r b_r}{6{\cdot}5 + 8\frac{a_r}{b_r} + 2{\cdot}5\frac{b_r}{a_r}} \tag{1.228}$$

In the case of restrained slabs, the influence of the corner levers can be taken into account by increasing m by 10%.

The transition from the previous formulae (1.221)-(1.224) to (1.228) takes place at values of a/b_r of approximately 0·5 to 0·6.

The reactions are determined in the same way as in (1.214), and it will immediately be seen that we get the same formula, but with ½ p instead of p, and $\bar{p}_b = 0$.

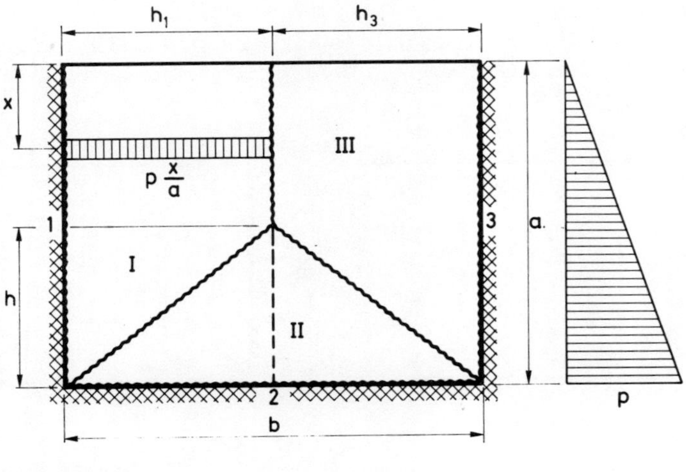

Fig. 1.222

1.23 Point load

The most dangerous position for a point load is at the free edge, a case that is dealt with in YLT, page 84. With this load, top reinforcement right out to the edge is required. With $m' = im$, we get, from (46) in YLT,

$$m = \frac{P}{2[\sqrt{i} + (1 + i) \arctan \sqrt{i}\,]} \tag{1.231}$$

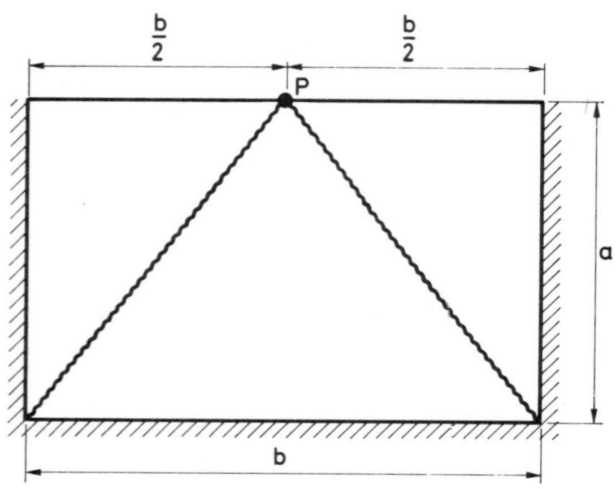

Fig. 1.23

In particular, $m' = m$, i.e. $i = 1$, means that $m = P/(2 + \pi) \sim P/5$. If the slab is not restrained, more dangerous yield patterns are found. Thus, for the pattern in Fig. 1.23, with unit deflexion at P, whereby the slab parts get the rotations $2/b$ and $1/a$, we get the work equation

$$P \times 1 = 2ma\,\frac{2}{b} + mb\,\frac{1}{a} = m\left(4\,\frac{a}{b} + \frac{b}{a}\right) \tag{1.232}$$

which, when $4 > b/a > 1$, gives a higher value of m than $P/5$, which was found above when $i = 1$. The highest value of m is $P/4$, which is obtained for $b = 2a$. However, point loading does not usually occur in the practical applications of this type of slab and is mostly of theoretical interest, so we will not give the matter more detailed attention here. It may be noted that the results in 1.4 are, to a certain extent, also valid for this type of slab.

1.24 Slabs with holes

These slabs can, for example, be dealt with by means of the work equation in the same way as we dealt with the slabs in 1.15. In the case of small holes with loading, the degrees of restraint are reduced as mentioned in section 1.15. Thus, in Fig. 1.241, the degree of restraint is reduced by c/a on side 1 and by d/b on side 2. In Fig. 1.242, the slab is regarded as half of a slab, which is supported on four sides and has two holes. On sides 1 and 3, the degrees of restraint are reduced by c/a, and on side 2, by d/b. As these slabs are hardly of much practical importance either, we will conclude with the calculation of the slab in Fig. 1.243.

When there is unit deflexion of the free edge, whereby all slab parts get the rotation $1/c$, the work equation is as follows:

$$4mc\,\frac{1}{c} = 2\left(\frac{1}{2}\,pac^2 - \frac{1}{3}\,pc^3\right)\frac{1}{c}$$

$$+ \left(\frac{1}{2}\,pbc^2 - \frac{2}{3}\,pc^3\right)\frac{1}{c}$$

$$+ 2 \times \frac{1}{2}\,\bar{p}c^2\,\frac{1}{c} + \bar{p}\,(2a + b - 4c) \times 1$$

or

$$m = \frac{P}{8} + \frac{\overline{P}}{4} - \frac{pc^2}{12} - \frac{\bar{p}c}{4} \tag{1.241}$$

where $P = pc(2a + b - 2c)$ and $\overline{P} = \bar{p}\,(2a + b - 2c)$. We further find that

$$m's = \frac{1}{6}\,psh^2 + \bar{p}(a - c + c)\,\frac{h}{2}$$

where $hs = c(a - c)$, $s^2 = c^2 + (a - c)^2$. Multiplication by s then gives

$$m' = \frac{p}{6}\,\frac{c^2(a - c)^2}{c^2 + (a - c)^2} + \frac{1}{2}\,\bar{p}a\,\frac{c(a - c)}{c^2 + (a - c)^2} \tag{1.242}$$

It is also possible to use formula (1.219) for the slab part at b and the corresponding formula with

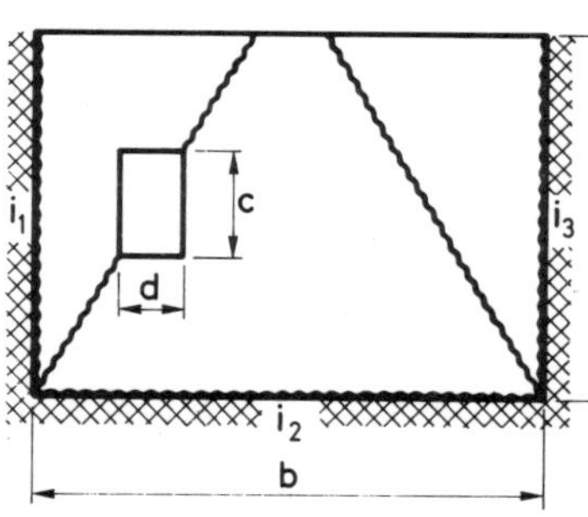

Fig. 1.241

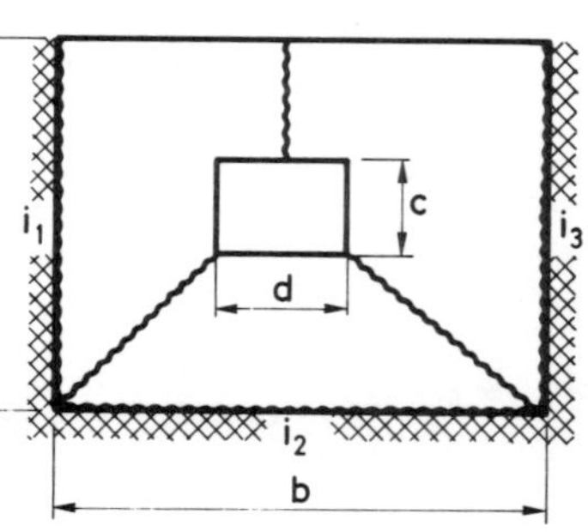

Fig. 1.242

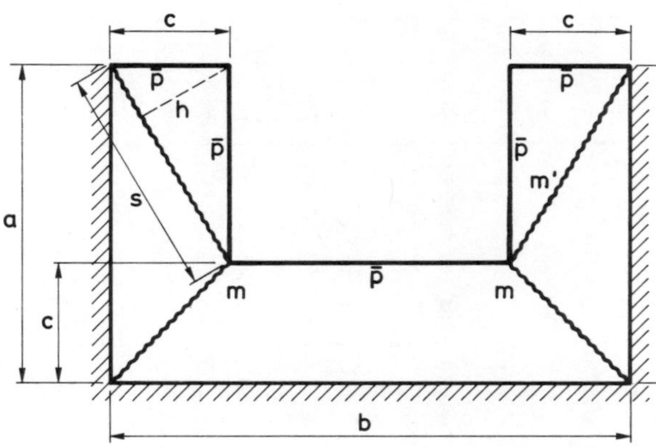

Fig. 1. 243

2 l for the slab part at a. (Regarding the latter formula, cf. the following section 1. 3.)

1. 25 Extent of top reinforcement

In the case of anchored slabs, which are only practicable in the case of simple support, i.e. when $i_1 = i_2 = i_3 = 0$, the top reinforcement corresponds to $m' = m$. With $b_r = b$, equations (1. 211) and (1. 212) give

$$m = \frac{p(1 + \alpha + 2\beta)^2 ab}{8(1 + \alpha + 2\beta) + 6(1 + 3\beta)\dfrac{a}{b}} \qquad (1.251)$$

The top reinforcement is carried out as shown in Fig. 1. 251 and must be taken out far enough for the panel with spans l and a, which has no top reinforcement, not to get a greater moment. This panel is calculated in exactly the same way as a simply supported slab that is not anchored, cf. (1. 217). Thus:

$$m \geqslant \frac{3p}{32(2 + 3\beta)}\left[\frac{4}{9}a^2 + (1 + \lambda + 2\beta)^2 l^2\right] \qquad (1.252)$$

where $\beta = \bar{p}_b/pa$, as usual, and $\lambda = \bar{p}_a/pl$. We can determine l from these two equations and then determine x by means of (1. 217), with l and λ instead of b and α.

For $\lambda = 0$ and $\beta = 0$, we get the following values for the ratios α_1 and α_2 shown in Fig. 1. 251:

a/b	α_1	α_2
0·1	0·25	0·42
0·2	0·17	0·38
0·3	0·11	0·34
0·4	0·07	0·30
0·5	0·04	0·25
0·6	0·02	0·19
0·7	0	0·13
0·8	—	—

When the load is so large that $\beta = 1$, we get, correspondingly, with $\lambda = 0$,

a/b	α_1	α_2
0·1	0·28	0·48
0·2	0·20	0·46
0·3	0·14	0·44
0·4	0·10	0·42
0·5	0·07	0·40
0·6	0·05	0·38
0·7	0·02	0·35
0·8	0·01	0·32

For smaller values of β, we interpolate between the two Tables.

Point loads on the free edge require a certain degree of top reinforcement, which has to be taken right out to the edge.

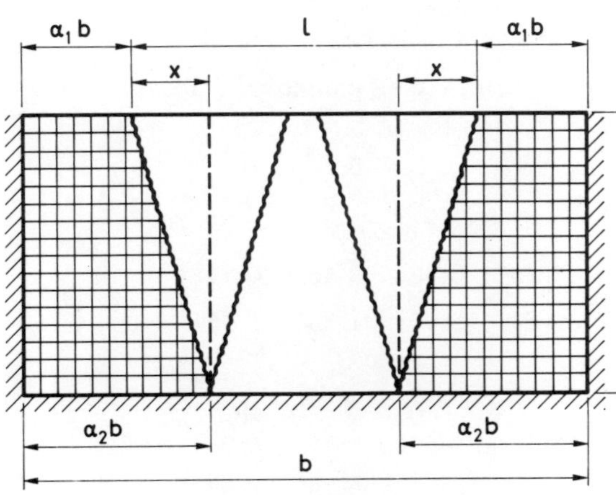

Fig. 1. 251

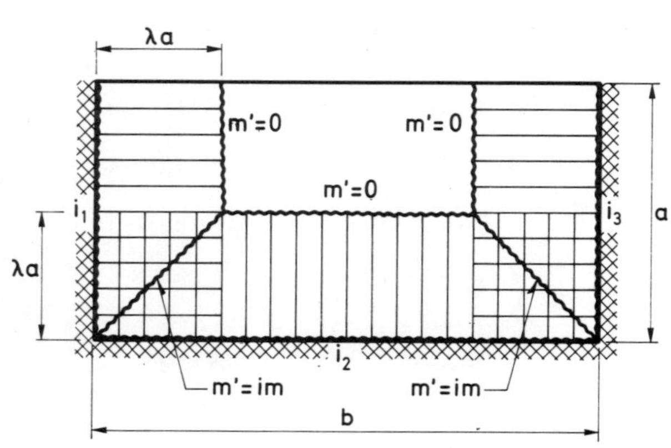

Fig. 1. 252

For the restrained slabs, which are calculated in accordance with (1.218) or (1.216), we have to determine how far the top reinforcement at the restrained edges should extend into the slab. We then assume that the slab is only loaded with its weight g and that adjacent slabs are loaded with the total load q. The yield pattern with negative moments shown in Fig. 1.252 is thus formed. It will immediately be seen by comparison with Fig. 1.16 that the equilibrium equations will be exactly the same as for the slab supported on four sides, with the side 2a and $i_4 = i_2$.

Formulae (1.161) to (1.163) are thus directly applicable, but not the following formulae as we now have a different expression for m. If the reinforcement is carried an equal distance in on all sides, and we denote this distance λa, then we obtain from (1.163), with $i_4 = i_2$, and 2a instead of a, $\lambda a = \alpha \times 2a$, i.e. $\lambda = 2\alpha$,

$$m(i_1 + 2i_2 + i_3 - 4\lambda i) = \lambda\left[1 - \left(\frac{1}{2} + \frac{a}{b}\right)\lambda + \frac{2}{3}\frac{a}{b}\lambda^2\right] gab \qquad (1.251')$$

where i_1, i_2 and i_3 are equal either to zero or to i. As m is determined by the total load q, the equation gives values for q/g.

These slabs are often used as flights of stairs, where $i_2 = 0$. Putting $i_1 = i_3 = i = 1$, formula (1.217), with $b_r = b/\sqrt{2}$, gives

$$\frac{q}{g} = \frac{192\lambda\left[1 - \left(\frac{1}{2} + \frac{a}{b}\right)\lambda + \frac{2}{3}\frac{a}{b}\lambda^2\right]}{(1 - 2\lambda)\left(8\frac{a}{b} + 9\frac{b}{a}\right)} \qquad (1.252'')$$

From this we find the following values for q/g:

	λ					
a/b	0·15	0·20	0·25	0·30	0·35	0·40
0·5	1·6	2·4	3·6	4·8	8·0	11·4
0·4	1·4	2·1	3·1	4·2	6·9	10·2
0·3	1·1	1·7	2·5	3·5	5·5	8·5
0·2	—	1·2	1·8	2·5	3·9	6·1

And further, in a corresponding manner, $i_1 = i_2 = i_3 = i = 1$ gives

$$\frac{q}{g} = \frac{1}{2}\frac{\lambda}{1-\lambda}\left(3{\cdot}5 + 12\frac{a}{b} + \frac{b}{a}\right) \times \left[1 - \left(\frac{1}{2} + \frac{a}{b}\right)\lambda + \frac{2}{3}\frac{a}{b}\lambda^2\right] \qquad (1.253)$$

from which we obtain the following values for q/g:

	λ				
a/b	1/4	1/3	1/2	2/3	3/4
0·5	1·6	2·0	3·4	5·5	7·5
0·4	1·5	2·0	3·3	5·6	7·7
0·3	1·5	2·0	3·4	5·8	8·0
0·2	1·6	2·1	3·7	6·5	9·0

The middle part of the panel, with the sides $a - \lambda a$ and $b - 2\lambda a$, can be regarded as a simply supported, unanchored slab. The top reinforcement must therefore be carried out far enough for this panel not to get a greater moment than the whole slab, i.e., from (1.217):

$$m \geqslant \frac{3qa^2}{64}\left[\frac{4}{9}(1-\lambda)^2 + \left(\frac{b}{a} - 2\lambda\right)^2\right] \qquad (1.254)$$

from which λ can be determined. Here, the limit of the top reinforcement is placed obliquely (Fig. 1.253), and λ_1 is determined by means of x in (1.217). We find

$$\lambda_1 = \frac{3}{8}\frac{b}{a} + \frac{\lambda}{4} - \frac{1}{2}\frac{(1-\lambda)^2}{\frac{b}{a} - 2\lambda} \qquad (1.255)$$

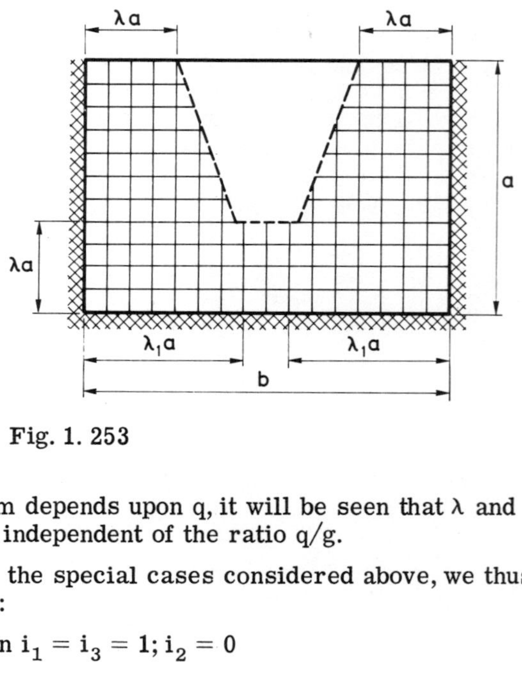

Fig. 1.253

As m depends upon q, it will be seen that λ and λ_1 are independent of the ratio q/g.

For the special cases considered above, we thus find:

when $i_1 = i_3 = 1$; $i_2 = 0$

a/b =	0·5	0·4	0·3	0·2
λ =	0·26	0·33	0·46	0·71
λ_1 =	0·63	0·90	1·30	2·05

when $i_1 = i_2 = i_3 = 1$

a/b =	0·5	0·4	0·3	0·2
λ =	0·35	0·48	0·75	(>1)
λ_1 =	0·68	0·97	1·42	—

These are normally the decisive values. The slabs with a >b are treated as half slabs with four supported sides.

1.26 Deflexions

The maximum deflexion obviously occurs at the middle of the free edge, so the strip must be placed along this edge, and $l = b$, (1.17). If it is now required that $u \leqslant b/500$, the thickness must be increased considerably for slabs for which the free edge is the longest side, whilst where the opposite is true, we get the same conditions as in slabs with four supported sides. When the slab thickness is large, the degree of reinforcement will be low, and it it may then be too low. The limit can be said to be reached when the yield moment is equal to the ultimate bending moment of the concrete cross-section, i.e. equal to $\frac{1}{6}\sigma h^2$. As we usually have a factor of safety of 2 against yielding, $m_y = 2m$, i.e. the limit is reached when $m = \frac{1}{12}\sigma h^2$. The bending strength σ of the concrete lies at about 40 kg/cm², and if we insert the slightly generous value of 48 kg/cm², we get $m = 4h^2$, i.e. $h = 0{\cdot}5\sqrt{m}$. Lower values of σ result in higher values of h, i.e. more favourable conditions since this is an upper bound for h. If a higher value of h than this is required on account of the stiffness, the reinforcement should not be calculated by means of m but by means of $4h^2$. Such slabs are, of course, not very economical and should be avoided, not only for reasons of economy, but also because the assumptions of the yield-line theory regarding a well developed plastic collapse state can be called into dispute. The tests carried out do not cover this case either. These considerations apply quite generally and therefore give

$$h \leqslant 0{\cdot}5\sqrt{m} \qquad (1.26)$$

as the limit for the total thickness of the slab. Finally, we will compare the stiffness of the simply supported, anchored slab with uniformly distributed load with that of the corresponding cantilever slab. For the former slab, (1.212) gives m and (1.17) gives u. For the cantilever slab with span a, we get from elementary beam theory, $u = pa^4/8EI$. The two slabs get the same m when $pa^4 = mb^2$, which gives $8 + 6(a/b) = (b/a)^3$; $b = 2{\cdot}2a$. Slabs with a lower value of b thus have a greater stiffness than the corresponding cantilever slabs.

The permanent deflexions due to creep and shrinkage can be estimated as explained at the end of section 1.17, with $l = b$.

1.27 Numerical example

A square water tank, measuring 4 × 4 × 2 m, is constructed from reinforced concrete slabs. The walls are free at the top and are restrained at the bottom and by each other. With the degrees of restraint $i_1 = i_2 = i_3 = 1$ and $p = 2000$ kg/m², equation (1.224) gives $m = 240$ kg with $b_r = 4/\sqrt{2} = 2{\cdot}83$ and $a = 2{\cdot}0$. On the other hand, with the same values and $a_r = 2/\sqrt{2}$, equation (1.228) gives $m = 258$ kg. Because of the corner levers, this is increased by 10% to $m = 285$ kg. With this value of m and the stresses 60 and 1000 kg/cm², we get $h_n = 0{\cdot}289\sqrt{285} = 4{\cdot}9$ cm, $f = 0{\cdot}410\sqrt{285} = 6{\cdot}9$ cm² per m. With $h = 8$ cm, we then find the maximum deflexion,

$$u = \frac{285 \times 400^2}{8 \times 200\,000 \times \frac{1}{12} \times 8^3} = 0{\cdot}67 \text{ cm}$$

$$< \frac{400}{500} = 0{\cdot}8 \text{ cm}$$

so the stiffness requirements are fulfilled.

The influence of the corner levers is neglected when calculating the reactions, so formula (1.214) is used with $\frac{1}{2}p = 1000$ kg/m² instead of p, as mentioned at the end of 1.22. From this, we find

$$\bar{q}_2 = 1000 \times 2 - \frac{8 \times 285 \times 2}{8} = 1430 \text{ kg/m}$$

$$H_{1,2} = H_{2,3} = 4m = 1140 \text{ kg}$$

$$Q_1 = Q_3 = 2 \times 285\left(\sqrt{2} - 1 + 2\frac{2}{2\sqrt{2}}\right)\sqrt{2} = 1475 \text{ kg}$$

$$R_1 = R_3 = 2\sqrt{2} \times 285 = 805 \text{ kg}$$

Check: reactions = load

$$4 \times 1430 + 2 \times 1475 + 2 \times 805 - 2 \times 1140$$

$$= 1000 \times 2 \times 4$$

The reactions q_2 call for tensile reinforcement in the bottom slab, Q in the side walls, and R at the free edge. The forces are so small that they do not affect the above design.

1.3 Supported along two adjacent sides

1.31 Uniformly distributed load and line load

This type of slab can immediately be related to the foregoing by imagining a symmetrically located and corresponding slab added, as shown in Fig. 1.31. We thus get a slab that is supported along three sides and free on the fourth side. This slab has the side lengths a and 2b and the loads p, $\bar{p}_b$ and $2\bar{p}_a$, so $\alpha = 2\bar{p}_a/2pb = \bar{p}_a/pb$ and $\beta = \bar{p}_b/pa$ are the same for both slabs. That this way of treating the problem is

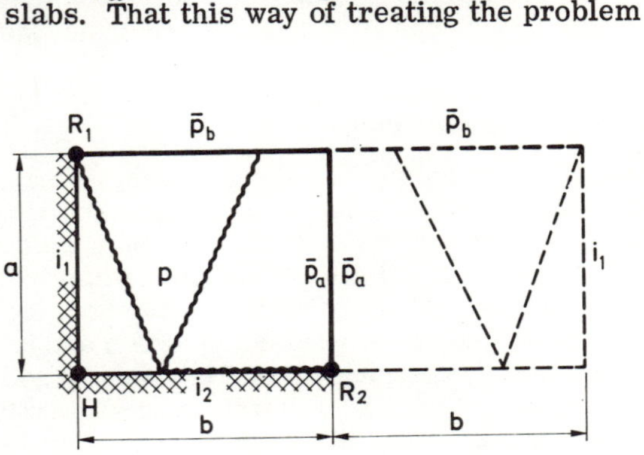

Fig. 1.31

correct can be seen in the same way as with Fig. 1. 212. This assumes that the yield line intersects the free edge b. If this is not the case, we have instead to consider the slab with the sides 2a and b and the loads p, $\bar{p}_a$ and $2\bar{p}_b$.

Assuming the conditions in Fig. 1. 31, we obtain from (1. 212) and (1. 218) respectively

$$m = \frac{pa'b'}{4 + 1{\cdot}5 \frac{a'}{b'}} \text{ for } i_1 = i_2 = 0 \qquad (1.311)$$

where a′ and b′ are determined from (1. 211), and

$$m = \frac{pab}{1{\cdot}5\sqrt{1+i_1} + 3(1+i_1)\frac{a}{b} + i_2\left(\sqrt{1+i_1} + \frac{2b}{a}\right)} \qquad (1.312)$$

where $b_r = b/\sqrt{1+i_1}$ has been used. This only applies when $\alpha = \beta = 0$; when this is not the case, (1. 216) must be used, with $b_r = 2b/\sqrt{1+i_1}$, which gives

$$k^2\left[9(1+i_2)(1+\alpha+2\beta)^2 + (1+i_1)\frac{a^2}{b^2}\right] - 2k\left[3(1+\alpha+2\beta)(1+i_2)i_2 + (1+i_1)\left(4+3i_2 + 6\beta(1+i_2)\frac{a^2}{b^2}\right)\right] + i_2^2\left[1+i_2+(1+i_1)\frac{a^2}{b^2}\right] = 0 \qquad (1.313)$$

where $k = pa^2/6m$, i.e. $m = pa^2/6k$.

In the diagonal section from R_1 to R_2, we find the negative moment

$$m' = \frac{pab + 3(\bar{p}_a a + \bar{p}_b b)}{6\left(\frac{a}{b}+\frac{b}{a}\right)} = \frac{P + 3\bar{P}}{6\left(\frac{a}{b}+\frac{b}{a}\right)} \qquad (1.314)$$

which is derived in the same way as (1. 242). From the diagonal section, the negative moment increases towards the corner, to m in simply supported slabs and to the restraining moment in restrained slabs. There are thus negative moments throughout the slab, so top reinforcement has to be provided throughout the slab, although it does not have to be equally heavy everywhere.

The influence of the corner levers is neglected when determining the reactions. A comparison between the slab supported on two sides and that supported on three sides (Fig. 1. 31) then shows immediately that the reactions at side a are the same for both slabs. On the other hand, the reactions on side b are no longer uniformly distributed as a consequence of the free edge, but get instead a distribution similar to that at a, i.e. a concentrated load $R_2 = 2m\sqrt{1+i_2}$ at the edge and Q_2 distributed in the same way as Q_1. As the reactions on a are known, and the sum must equal the load on the slab, the reactions on b can easily be found. We now get from (1. 214), with $2b/\sqrt{1+i_1}$ instead of b_r:

$$Q_1 = 2m\left(\sqrt{1+i_2} - 1 + \sqrt{1+i_1}\,\frac{a}{b}\right)\sqrt{1+i_1}$$

$$R_1 = 2m\sqrt{1+i_1};\ Q_2 = pab + \bar{p}_a a + \bar{p}_b b - 2m\left(\sqrt{1+i_2} + (1+i_1)\frac{a}{b}\right);\ R_2 = 2m\sqrt{1+i_2}$$

$$H = 2m\sqrt{1+i_1}\sqrt{1+i_2} \qquad (1.315)$$

where the latter is negative. It will be seen that the sum is equal to the load.

1. 32 Point load

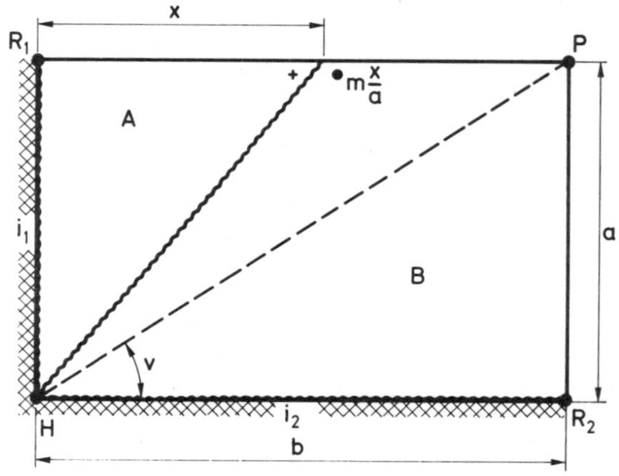

Fig. 1. 32

The most dangerous location is obviously at the free corner. According to YLT, page 85, it is then required that

$$m = m' = \tfrac{1}{2}P \qquad (1.321)$$

The yield pattern shown in Fig. 1. 32 gives the nodal force $m(x/a)$ at the free edge. The moment equation for A gives

$$m(1+i_1)a = m\frac{x}{a}x;\quad \frac{x}{a} = \sqrt{1+i_1}$$

For B, we get

$$mi_2b + mx = Pa - m\frac{x}{a}a$$

from which we get, by inserting x,

$$m = \frac{P}{2\sqrt{1+i_1} + \frac{b}{a}i_2} \qquad (1.322)$$

which only yields the same as (1. 321) in the case of simple support, but otherwise gives less, so (1. 321) is always valid. Restraint thus has no effect. It may be noted that with simple support we get $R_1 = R_2 =$

$H = 2m = P$, where H acts in the opposite direction to R_1 and R_2. $Q_1 = Q_2 = 0$. Thus, we only get reactions at the corners, and it will immediately be seen that in this case we have one of the classic solutions from the theory of elasticity. The deflexion surface is

$$z = \frac{Pxy}{2EI} \quad (1.323)$$

and the slab is subjected to pure torsion.

1.33 Deflexions

As the deflexion surface has horizontal tangents at the supports, it has two horizontal tangents at H, i.e. a horizontal tangent plane. A rectilinear strip from H to the free corner can therefore be regarded as a beam clamped at H. The maximum moment occurs at H and is m′ according to the yield-line theory and generally slightly less than according to the theory of elasticity, which approximately applies at the stage at which the deflexions are calculated. Analogously with 1.17, the maximum deflexion u, which occurs at the free corner, is estimated by using the value m′ and by assuming that the moment in the strip decreases linearly to zero at the edge. Hence,

$$EIu = \tfrac{1}{3}m'l^2 = \tfrac{1}{3}m'(a^2 + b^2) \quad (1.331)$$

Because the corner is anchored we must always have $m' = m$ at the corner, whether the slab is restrained or not. Therefore,

$$EIu \sim \tfrac{1}{3}m(a^2 + b^2) \quad (1.332)$$

For simply supported slabs, u can incidentally be calculated correctly. Since a force 1 acting at the free corner gives the deflexion surface $z_1 = xy/2EI$, an arbitrary load p will, according to Betti's theorem $\Sigma P_k z_i = \Sigma P_i z_k$, give the deflexion u at the corner,

$$u \times 1 = \iint \frac{pxy}{2EI}\,dx\,dy \quad (1.333)$$

Thus we get $EIu = \tfrac{1}{8}pa^2b^2$ for a uniformly distributed load p, while according to (1.331)

$$EIu = \frac{pab(a^2 + b^2)}{12 + 4{\cdot}5\frac{a}{b}}$$

which agrees well with the correct value, provided the lengths of the sides do not differ too much.

The contribution from the line loads in (1.333) becomes $EIu = \tfrac{1}{4}(\bar{p}_a a + \bar{p}_b b)ab$, i.e. a total from p, $\bar{p}_a$ and $\bar{p}_b$ of

$$u = \frac{pab + 2\,\bar{p}_a a + 2\,\bar{p}_b b}{8EI}\,ab = \frac{P + 2\bar{P}}{8EI}\,ab \quad (1.334)$$

Statically and elastically, a restrained beam may be said to correspond to half a beam supported at both ends, so a suitable requirement for stiffness would be as follows

$$u \leqslant \frac{2l}{500} = \frac{\sqrt{a^2 + b^2}}{250}$$

This type of simply supported slab is naturally highly flexible. It is of some interest to compare it with a cantilever slab of length c. With a uniformly distributed load the two slabs get the same deflexion when $c^2 = ab$. A square slab with two simply supported sides thus has the same stiffness as the corresponding cantilever slab.

1.34 Numerical example

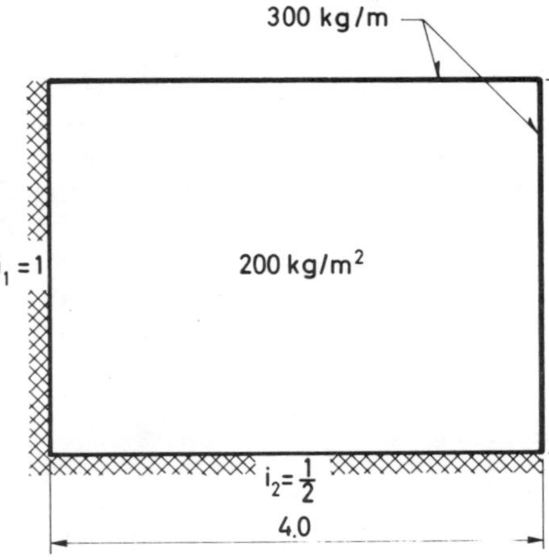

Fig. 1.34

An oriel in a house is to be supported by a reinforced concrete slab. The dimensions in metres and the load are as shown in Fig. 1.34. The degrees of restraint are ½ and 1. The deadweight is estimated to be 300 kg/m², so p = 500 kg/m². The ratios between the loads are $300/(500 \times 3) = 0{\cdot}2$ and $300/(500 \times 4) = 0{\cdot}15$. Formula (1.313) must be used, but as the restraints are unequal, it cannot be decided in advance which of the sides is to be inserted as a or b. Both possibilities must therefore be investigated.

With $a = 3{\cdot}0$, $b = 4{\cdot}0$, we get $\alpha = 0{\cdot}15$, $\beta = 0{\cdot}20$, $i_1 = 1$ and $i_2 = 0{\cdot}5$.

Thence,

$$k^2[9 \times 1{\cdot}5 \times 1{\cdot}55^2 + 2 \times 0{\cdot}75^2] - 2k[3 \times 1{\cdot}55 \times 1{\cdot}5 \times 0{\cdot}5 + 2(5{\cdot}5 + 1{\cdot}2 \times 1{\cdot}5 \times 0{\cdot}75^2)] + 0{\cdot}25(1{\cdot}5 + 2 \times 0{\cdot}75^2) = 0$$

$$33{\cdot}5k^2 - 2k \times 16{\cdot}5 + 0{\cdot}66 = 0;\ k = 0{\cdot}967$$

$$m = \frac{500 \times 9}{6 \times 0{\cdot}967} = 775 \text{ kg}$$

On the other hand, with $a = 4{\cdot}0$, $b = 3{\cdot}0$, $\alpha = 0{\cdot}2$, $\beta = 0{\cdot}15$, $i_1 = 0{\cdot}5$ and $i_2 = 1$, we get

$$k^2[9 \times 2 \times 1{\cdot}5^2 + 1{\cdot}5 \times 1{\cdot}33^2] - 2k[3 \times 1{\cdot}5 \times 2 + 1{\cdot}5(7 + 0{\cdot}9 \times 2 \times 1{\cdot}33^2)] + 2 + 1{\cdot}5 \times 1{\cdot}33^2 = 0$$

$$43{\cdot}2k^2 - 2k \times 24{\cdot}3 + 4{\cdot}67 = 0;\ k = 1{\cdot}017$$

$$m = \frac{500 \times 16}{6 \times 1{\cdot}017} = 1310 \text{ kg}$$

With the stresses 75 and 1300 kg/cm^2, we therefore get $h_n = 0{\cdot}260\sqrt{1310} = 9{\cdot}4$ cm and $f = 0{\cdot}348\sqrt{1310} = 12{\cdot}6$ cm^2 per m, h = 13 cm. As $m' = 1310(16 + \frac{1}{2} \times 9)/25 = 1070$ kg, the deflexion will be

$$u = \frac{1070 \times 500^2}{3 \times 200\,000 \times \frac{1}{12} \times 13^3} = 2{\cdot}4 \text{ cm}$$

$$> \frac{500}{250} = 2{\cdot}0 \text{ cm}$$

The thickness of the slab should therefore be increased to 14 cm, whereby $u = 1{\cdot}9$ cm. In the diagonal section, the negative moment will be $(6000 + 3 \times 2100)/(6 \times 2{\cdot}08) = 1000$ kg. In this section, the top reinforcement required at the restrained edges gives only $1310\,[(\frac{1}{2} \times 16 + 9)/25] = 900$ kg, so a little extra reinforcement is necessary at the top.

1.4 Supported along two opposite sides

1.41 Uniformly distributed load and line load

As the treatment of uniformly distributed load and line load parallel with the supports is trivial, we will only investigate a line load acting perpendicular to the supports. The case in which the line load is the only load has already been dealt with in YLT, example 30, page 117, for simple supports. When l is the span, the result can be represented by means of the approximate formula

$$m = \frac{\bar{p}l}{12 + 7i}; \quad i = \frac{m'}{m} \qquad (1.411)$$

If there are the restraining moments m′ at both supports, we obviously get $m + m' = \frac{1}{12}\bar{p}l$, as the yield pattern is the same as for i = 0, but the internal moments are now given by m + m′ instead of by m (cf. example 8 in YLT, page 24). With a single restraint, we take the mean value between 0 and 2 restraints. This results in the following formula for restrained slabs:

$$m = \frac{\bar{p}l}{12 + (7 + 2{\cdot}5n)i} \qquad (1.412)$$

where n is the number of restraints, i.e. n = 0, 1 or 2.

Taking the width of the slab as b and calculating the slab as a beam, we get

$$\frac{1}{8}\frac{\bar{p}l^2}{b} = m - \frac{1}{2}(m_1' + m_3')$$

or, since m_1' and m_3' are equal either to m′ or to zero,

$$m = \frac{\bar{p}l^2}{(8 + 4ni)b} \qquad (1.413)$$

This formula must be used when it gives a higher value of m than the above two formulae.

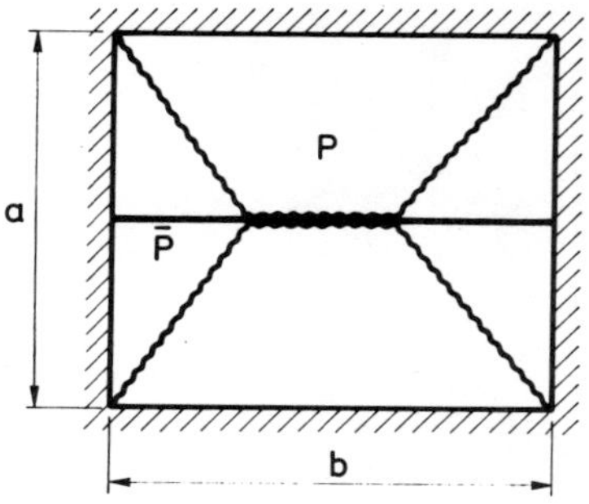

Fig. 1.411

It may be of some interest to compare the results in 1.13 with those obtained by superposition. For a square slab, simply supported on all four sides and subjected to a uniformly distributed load P and a uniformly distributed line load $\bar{P}$, superposition gives

$$m = \frac{P}{24} + \frac{\bar{P}}{12} = \frac{P}{24}(1 + 2\beta), \quad \bar{P}/P = \beta$$

On the other hand, from the transformations (1.136) and (1.137), with $\alpha = 0$ and (1.116″), we get

$$m = \frac{P}{8}\,\frac{(1 + 2\beta)^2}{2 + 5\beta + \sqrt{(1 + 2\beta)(1 + 3\beta)}}$$

The fact that it is α and not β that is to be put at zero can be seen from Fig. 1.411, since b is the side that is parallel with the middle yield line (cf. Fig. 1.111).

The two formulae yield the following values:

β	P/m by superposition	P/m by combination
0	24	24
1/4	16	17·3
1/2	12	13·5
3/4	9·6	11·0
1	8	9·3

It is thus up to 16% on the safe side to use superposition for these values of β. Using superposition in the case of a corresponding slab with b = 2a, we get

$$m = \frac{P}{28} + \frac{\bar{P}}{12}$$

When $\bar{P}$ is parallel with b, the formulae in 1.13, with $\alpha = 0$, give

$$m = \frac{P}{8}\,\frac{(1 + 2\beta)^2}{2{\cdot}5 + 5{\cdot}5\beta + \sqrt{(1 + 2\beta)(1 + 3\beta)}}$$

and when $\bar{P}$ is parallel with a,

$$m = \frac{P}{8}\,\frac{(1 + \alpha)^2}{2{\cdot}5 + 2\alpha + \sqrt{1 + \alpha}}$$

From this we get the following values:

$\overline{P}/P$	P/m by superposition	P/m by combination for $\overline{P} \dotplus b$	P/m by combination for $\overline{P} \dotplus a$
0	28	28	28
1/4	17·7	19·6	21·1
1/2	12·9	15·0	16·8
3/4	10·2	12·1	13·9
1	8·4	10·2	11·8

which shows that the use of superposition here is considerably more on the safe side, so it pays to use the formulae in 1.13.

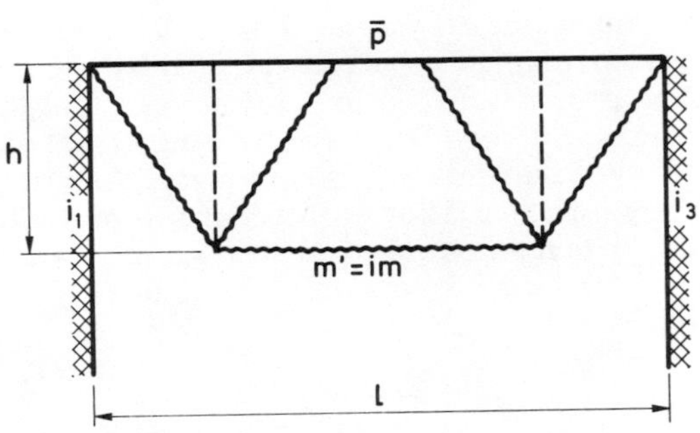

Fig. 1.412

When the line load acts on the free edge, we can, as an approximation, use the yield pattern shown in Fig. 1.412, which will be seen to be the same as that in Fig. 1.215, with b = l, a = h and $i_2 = i$. According to the maximum principle, h is determined so that the moment is maximum. This is determined from (1.216), with $\alpha = 0$.

When $p \to 0$, $\beta \to \infty$ but $k\beta \to \bar{p}a/6m$. Introducing $K = \bar{p}b_r/m = 6(b_r/a)k\beta$, we get

$$K^2 - 2K\left(i\frac{l_r}{h} + 4\frac{h}{l_r}\right) + i^2\left(\frac{4}{1+i} + \frac{l_r^2}{h^2}\right) = 0 \qquad (1.414)$$

when a and b_r are replaced by h and l_r, and $l_r = 2l/(\sqrt{1+i_1} + \sqrt{1+i_3})$.

The equation can naturally also be developed from the original equilibrium equations with p and $\bar{p}_a$ equal to zero. If we introduce $h = zl_r$ for the sake of brevity, the maximum principle gives $dK/dz = 0$, and differentiation of (1.414) with regard to z yields

$$-2K\left(-\frac{i}{z^2} + 4\right) + i^2\left(-\frac{2}{z^3}\right) = 0$$

or

$$K = \frac{i^2}{z(i - 4z^2)} \qquad (1.415)$$

By inserting this in (1.414) we get

$$16z^4 + 4(3 + i)z^2 = i(2 + i)$$

from which z can easily be calculated, and thereby also K, which in turn gives m.

It will be seen from (1.415) that i = 0 gives K = 0, i.e. $m = \infty$ or $\bar{p} = 0$.

In other words, a line load only, acting on a free edge, calls for top reinforcement right out to the edges just as in the case of point loads. We further find the following values:

i	z	$\bar{p}l_r/m = K$
1/4	0·203	3·6
1/2	0·286	5·05
3/4	0·349	6·15
1	0·401	7·0
2	0·565	9·8
3	0·689	11·9
4	0·794	13·6

We see that $K \sim 7\sqrt{i}$ and thereby find the approximate formula

$$m = \frac{\bar{p}l_r}{7\sqrt{i}} \qquad (1.416)$$

or

$$\sqrt{mm'} = \tfrac{1}{7}\,\bar{p}l_r$$

which shows that neither m nor m′ can be equal to zero. The required amount of reinforcement can be taken as proportional to $m + m' = m(1 + i) = \tfrac{1}{7}\,\bar{p}l_r\,[\sqrt{i} + \sqrt{(1/i)}]$, which is minimum for i = 1. The most economical reinforcement is thus obtained at

$$m = m' = \tfrac{1}{7}\,\bar{p}l_r \qquad (1.417)$$

Here, too, we can investigate the use of superposition for the slabs dealt with in section 1.21. With $i_1 = i_3 = 0$ and $i_2 = 1$, we get, for example for b = 2a from (1.218) and (1.416):

$$m = \frac{P}{15} + \frac{\overline{P}}{7}$$

For $\beta = 1$, i.e. $P = \overline{P}$, we get m = P/4·77, whilst (1.216) gives k = 0·409 and m = P/4·90. With b = 3a and the remaining values unaltered, superposition yields the same values as above, whereas combination now gives k = 0·270 and m = P/4·86. Thus, in these cases and, of course, for lower values of $\overline{P}$, we would do well to use the far simpler calculation achieved by superposition.

When $i_2 = 0$, superposition becomes meaningless, as (1.416) no longer gives finite values. If, for example, we attempt to use (1.416) with i = 1 in the case of an anchored slab with b = 2a and $\overline{P} = P$, superposition using (1.212) gives $m = (P/11) + (\overline{P}/7) = P/4{\cdot}28$, whilst combination by means of (1.211) and (1.212) gives $m = P/4$, i.e. a slightly higher value of m. For $b = 3a$ we get, correspondingly, $P/4{\cdot}11$ and $P/3{\cdot}56$.

1.42 Point load

For a point load acting on the free edge, formula (1.231) can be applied unaltered. Reinforcement is required at the top right out to the edge. For a point load acting at a considerable distance from the edge, (1.141) and (1.142) apply, and reinforcement is not necessary at the top. In the following we will study the transition from one case to the other, i.e. a point load acting near the free edge, and will, in particular, determine how far from the edge the load has to be located to render top reinforcement unnecessary.

Let the load act at a point at a distance c from the edge. Around the load we can expect a circular yield line, which possibly continues in two tangents. With the yield pattern shown in Fig. 1.421, all slab parts get the same rotation, namely, $1/r$ when P gets unit deflexion. The work equation then gives

$$P \times 1 = (m + m')\, 2\alpha r \frac{1}{r} + 2mc \sin\alpha \frac{1}{r} + 2\, m'\left(\frac{c}{\sin\alpha} + r \cot\alpha\right)\frac{1}{r}$$

$$P = (m + m')\, 2\alpha + 2m' \cot\alpha + 2\left(m.\sin\alpha + \frac{m'}{\sin\alpha}\right)\frac{c}{r}$$

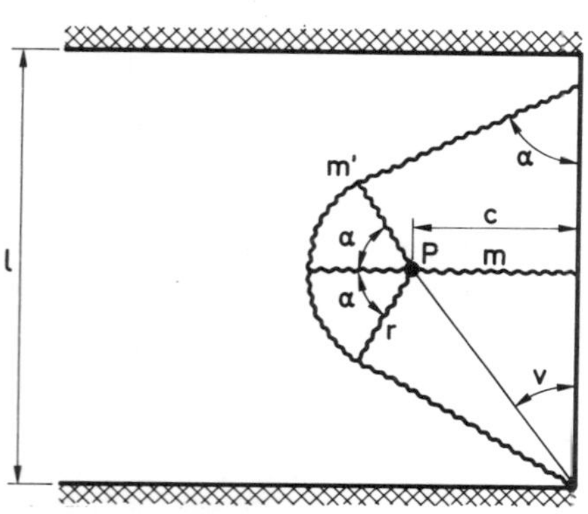

Fig. 1.421

According to the maximum principle, m must be maximum or P minimum. This means that r has to be as large as possible, i.e. the yield pattern must be such that the yield lines reach the corners of the slab. In this case, $r = (l/2)\sin\alpha - c\cos\alpha$. By inserting the angle v defined by $\tan v = 2c/l$, we get

$$P = 2(m + m')\alpha + 2m' \cot\alpha + 2\left(m \sin\alpha + \frac{m'}{\sin\alpha}\right)\frac{\sin v}{\sin(\alpha - v)}$$

$$= 2(m + m')\alpha + 2(m + m')\cot(\alpha - v) - 2m\frac{\cos\alpha \cos v}{\sin(\alpha - v)} \qquad (1.421)$$

where α is determined by means of $dP/d\alpha = 0$. This gives

$$\cos(\alpha - v) = \sqrt{\frac{m}{m + m'}}\cos v \qquad (1.422)$$

for determination of α. As $\alpha \leqslant \pi/2$, we must have $\tan v \leqslant \sqrt{m/(m + m')}$.

With this value of α, we get from (1.421)

$$P = 2(m + m')\left[v + \arccos\sqrt{\frac{m}{m + m'}}\cos v\right] + 2\cos v\left[\sqrt{m(m' + m\sin^2 v)} + m \sin v\right] \qquad (1.423)$$

For $m' = 0$ we get the special case of $\alpha = 2v$; $P = 2m(2v + \sin 2v)$. This expression shows that, when $v > 0$, the top reinforcement can be omitted, but for small values of v, we get such large moments that it would be uneconomical to omit top reinforcement. As the reinforcement required can usually be considered as proportional to $m + m'$, we can find the most economical reinforcement by choosing such a ratio of m'/m that $m + m'$ is minimum. As $d(m + m') = 0$, by differentiation with regard to m, (1.421) immediately gives $\cos\alpha = 0$, $\alpha = \pi/2$, $\tan v = \sqrt{m/(m + m')}$. With these values we obtain

$$m + m' = \frac{P}{\pi + 4\frac{c}{l}};\quad \frac{m}{m + m'} = 4\frac{c^2}{l^2} \leqslant 1 \qquad (1.424)$$

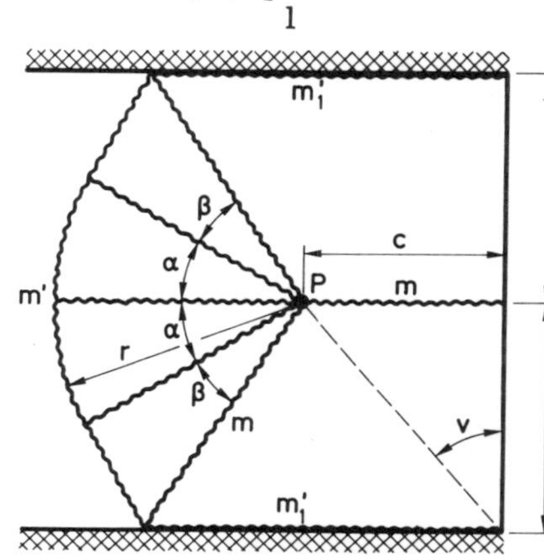

Fig. 1.422

It will be noted that the restraint conditions of the slab do not appear in the formulae. In order to include these, we must let the yield pattern intersect the supports, whereby we get the yield pattern shown in Fig. 1.422, in which we have assumed the same conditions at both supports, thereby obtaining a symmetrical yield pattern.

We apply the work equation with unit deflexion at point P.

We then get

$$P \times 1 = (m + m')r \times 2\alpha\frac{1}{r} + 2(m + m')\, r \tan\beta\frac{1}{r} + 2(m + m_1')\left(c + \frac{l}{2}\cot(\alpha + \beta)\right)\frac{2}{l}$$

Here, α and β are determined by means of the maximum principle.

$$\frac{\partial P}{\partial \alpha} = 0 \text{ gives: } m + m' = \frac{m + m_1'}{\sin^2(\alpha + \beta)}$$

and

$$\frac{\partial P}{\partial \beta} = 0 \text{ gives: } \frac{m + m'}{\cos^2 \beta} = \frac{m + m_1'}{\sin^2(\alpha + \beta)}, \text{ i.e. } \beta = 0$$

$$\sin \alpha = \sqrt{\frac{m + m_1'}{m + m'}} \text{ or } \cot \alpha = \sqrt{\frac{m' - m_1'}{m + m_1'}}$$

With these values we find that

$$P = 2(m + m') \operatorname{arccot} \sqrt{\frac{m' - m_1'}{m + m_1'}} + (m + m_1')\, 4\frac{c}{l} + 2\sqrt{(m' - m_1')(m + m_1')} \qquad (1.425)$$

We can assume that $m_1' = m'$ for the restrained slab. Then $\alpha = \pi/2$ and (1.425) gives the same $m + m'$ as (1.424) but, in this case, for all values of m/m'. When $c > (\pi/4)\, l$, (1.142) applies, i.e. $P = 2\pi(m + m')$. When $c < \frac{1}{2} l \sqrt{m/(m + m')}$, m/m' can no longer be selected arbitrarily if an economical reinforcement is desired, but must be determined by $m/(m + m') = 4(c^2/l^2)$. Thus, for restrained slabs, and for $c > \frac{1}{2} l \sqrt{m/(m + m')}$,

$$m + m' = \frac{P}{2\pi} \text{ for } c \geq \frac{\pi}{4} l; \quad \text{with } m/m' \text{ arbitrary}$$

$$m + m' = \frac{P}{\pi + 4\frac{c}{l}} \text{ for } \frac{\pi}{4} l \geq c; \quad \text{with } m/m' \text{ arbitrary} \qquad (1.424')$$

Otherwise,

$$\frac{m}{m + m'} = 4\frac{c^2}{l^2}$$

For simply supported slabs, $m_1' = 0$. We then find from (1.425) that

$$P = 4m\frac{c}{l} + 2\sqrt{mm'} + 2(m + m') \operatorname{arccot} \sqrt{\frac{m'}{m}} \qquad (1.426)$$

If it is assumed, as above, that the reinforcement required is proportional to $m + m'$, the most economical reinforcement is found in a corresponding manner for $m' = 0$. We then again get (1.424), although now with $m' = 0$. Equation (1.424) with $m' = 0$ therefore applies for $(\pi/4) l \geq c \geq \frac{1}{2} l$, and (1.141) for $c \geq (\pi/4) l$, i.e., $P = 2\pi m$. When $c < \frac{1}{2} l$, m' must be > 0, so the two formulae are no longer comparable.

We shall now return to formulae (1.423) and (1.426), corresponding to the two yield patterns. In Fig. 1.423, the quantity $m + m'$ proportional to the reinforcement required is plotted as ordinate, and $m/(m + m')$ as abscissa. Formula (1.423) must give a curve like ABC with minimum for $m/(m + m') = 4(c^2/l^2)$, while formula (1.426) gives a curve like DE, with its lowest value at $m' = 0$. For any value of $m/(m + m')$, the valid formula is that which gives maximum $m + m'$, i.e. the curve in full line gives $m + m'$. It will immediately be seen that the most economical reinforcement corresponds to the point of intersection S for the two formulae, provided DE lies above B, which is the case here.

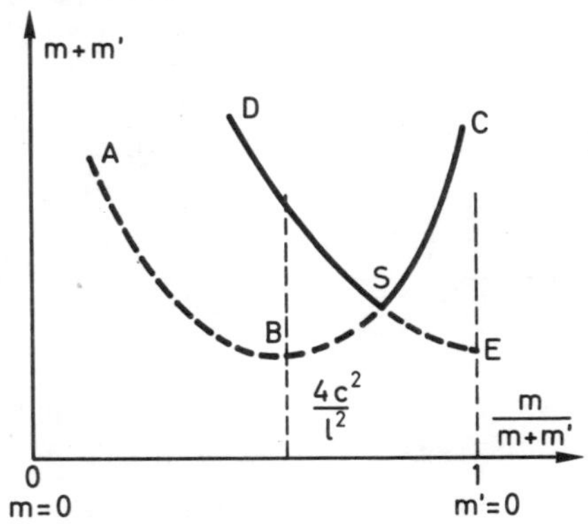

Fig. 1.423

With $\operatorname{arccot} \sqrt{m'/m} = \arcsin \sqrt{m/(m + m')}$, we find that

$$\frac{P}{2(m + m')} = v + \arccos \sqrt{\frac{m}{m + m'}} \cos v + \frac{m}{m + m'}\left(\sin v + \sqrt{\frac{m'}{m} + \sin^2 v}\right)\cos v$$

$$= \frac{m}{m + m'} \tan v + \frac{\sqrt{mm'}}{m + m'} + \arcsin \sqrt{\frac{m}{m + m'}} \qquad (1.427)$$

For the load on the edge we get $v = 0$, $m = m'$

$$m = m' = \frac{P}{2 + \pi}$$

When c varies from $\frac{1}{2} l$ to 0, m' will vary between 0 and m, while in both cases, m is $P/(2 + \pi) \sim P/5$. We further find, from solution of (1.427):

$\frac{m'}{m}$	$2c/l = \tan v$	P/m	$\frac{m}{m + m'}$	$\frac{1}{2}(1 + \tan v)$
0	1	5·14	1	1
¼	0·502	4·78	0·8	0·751
½	0·293	4·87	0·667	0·647
¾	0·132	5·00	0·571	0·566
1	0	5·14	0·5	0·5

It will be seen that we have, to a good approximation,

$$m = \frac{P}{5}, \quad \frac{m}{m + m'} = \frac{1}{2} + \frac{c}{l}, \quad \frac{c}{l} \leq \frac{1}{2} \tag{1.428}$$

As mentioned above, for higher values of c, the following applies:

$$m' = 0, m = \frac{P}{\pi + 4(c/l)}, c \leq \frac{\pi}{4} l, m = \frac{P}{2\pi}, c \geq \frac{\pi}{4} l \tag{1.424''}$$

It will be seen from this that the top reinforcement can be omitted as long as $c > \frac{1}{2} l$. In practice, it is usually preferable to avoid top reinforcement even though the total reinforcement has to be increased slightly. In this case it is necessary, as mentioned, to put

$$m = \frac{P}{2(2v + \sin 2v)} \sim \frac{P}{4\frac{c}{l}\left(4 - 3\frac{c}{l}\right)} \tag{1.423'}$$

which, however, gives an approximately 20% increase in the amount of steel, even at $c = \frac{1}{3} l$.

When the area over which the load is distributed is not too great, the contribution $P \times 1$ in the work equation changes to

$$\iint pz \, df = P \zeta$$

where z is the deflexion, and ζ its mean value over the loaded area. As the deflexion under the force resultant is 1, we get $\zeta = 1 - \frac{2}{3}(\rho/r)$, in the case of a conical surface of deflexion and $\zeta = 1 - \frac{1}{2}(\rho/r)$ in the case of a prismatic (roof-shaped) surface of deflexion, where ρ is the distance to the contour of the loaded area, and r the distance to the axis of rotation of the slab part in question. In the case of the figures under consideration here, we have as a rule $r = \frac{1}{2} l$, whilst the surface of deflexion is partly conical and partly roof-shaped. If the loaded area is rectangular, with the sides a and b, then $\pi\rho^2 \sim ab$, and

$$\zeta \sim 1 - \frac{2}{3}\frac{\sqrt{ab}}{l} \tag{1.429}$$

1.43 Wheel loads

We will next establish formulae for the case of two equal wheel loads P, the distance between the loads being a. When they are applied on the mid-line, we get the pattern shown in Fig. 1.431, which is immediately seen to give the same values as Fig. 1.422 with $c = \frac{1}{2}a$.

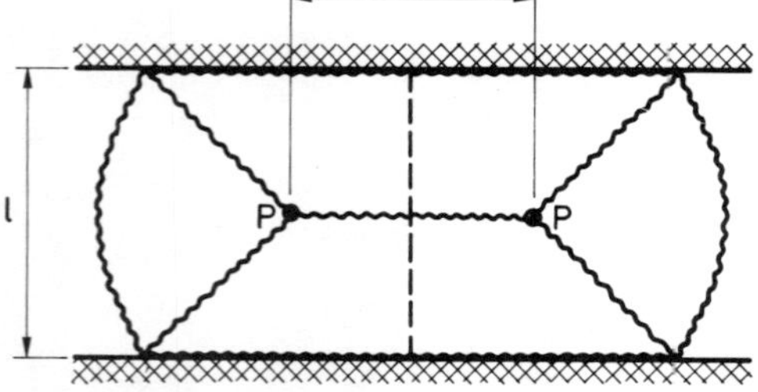

Fig. 1.431

If one of the wheel loads is applied near the edge, we get the yield patterns in Fig. 1.432, which correspond exactly to those dealt with in section 1.42. In the corresponding formulae, P must be replaced by 2P as the work done by the external forces is $P \times 1 + P \times 1 = 2P$.

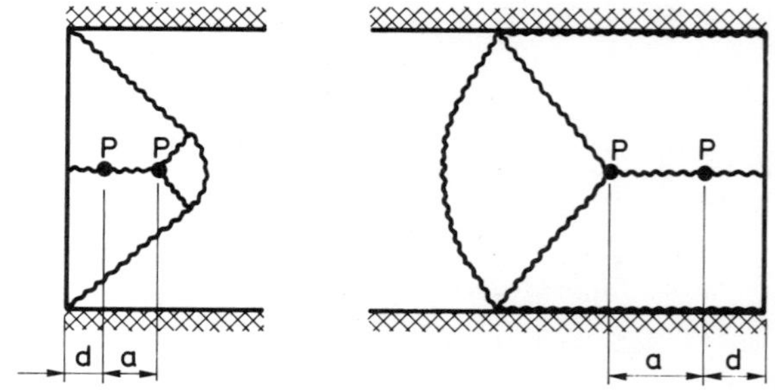

Fig. 1.432

The value of m calculated according to these formulae must, of course, not be lower than that corresponding to one of the point loads. Thus, corresponding to Fig. 1.431 with economical reinforcement ($m' = 0$), the following obtains for a simply supported slab:

$$m = \frac{P}{\pi + 2\frac{a}{l}} \geq \frac{P}{2\pi}; a \leq \frac{\pi}{2} l \tag{1.431}$$

For the restrained slab, m is replaced by $m + m'$. When $a > (\pi/2)l$, m or $m + m'$ is equal to $P/2\pi$.

When one load is applied near the free edge, we must utilize formulae (1.423) and (1.425), either with 2P and $c = a + d$ or with P and $c = d$, and it is the largest value that applies. The economical reinforcement is obtained at the "intersection" of these formulae in a similar manner to the case in Fig. 1.423. In practice, the simplest way of doing this is to use the formulae for several different values of m'/m. Four moments are obtained for each value, as both formulae have to be used in two ways, as mentioned above. It is then the greatest of these moments that is valid. In this way we get a series of values for various m'/m. The most economical of these is the one that gives the lowest value of $m' + m$.

When (1.144) is being used, the load applied furthest from the edge must not give a higher value of m than that stated above.

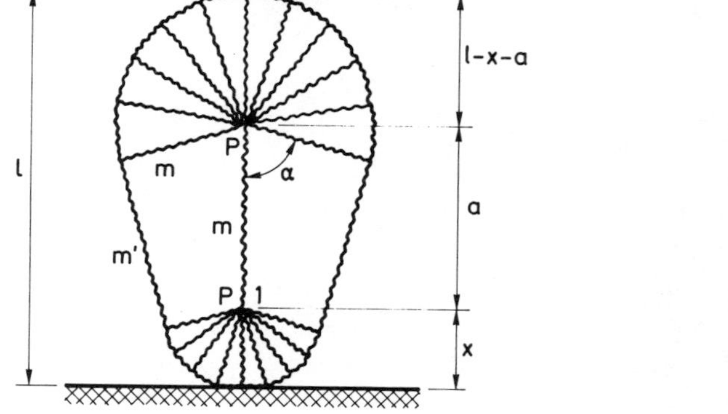

Fig. 1.433

When the line connecting the wheel loads is perpendicular to the supports, we can try the yield pattern in Fig. 1.433. If we limit ourselves to the case of simply supported slabs with $m' = 0$ and to slabs restrained on both sides with a support moment equal to m', the curved yield line will touch the edge (YLT, page 111 at top).

With the two circles shown, the work equation gives

$$P \times 1 + P \frac{1-a-x}{x} \times 1 = (m + m')2\alpha x \frac{1}{x}$$

$$+ (m + m')(2\pi - 2\alpha)(1 - a - x) \frac{1}{x}$$

$$+ 2(m + m')a \sin \alpha \frac{1}{x}$$

$$P \frac{1-a}{x} = 2(m + m')\left[2\alpha - \pi + (\pi - \alpha)\frac{1-a}{x} + \frac{a}{x} \sin \alpha\right]$$

with

$$\cos \alpha = \frac{1-a-2x}{a}, \quad x = \frac{1}{2}(1 - a - a \cos \alpha)$$

Introducing this, we get

$$P = (m + m')\left[\pi + \frac{2a}{1-a}\left\{\sin \alpha + \left(\frac{\pi}{2} - \alpha\right)\cos \alpha\right\}\right]$$

The most dangerous position corresponds to $dP/d\alpha = 0$, which immediately gives $\alpha = \pi/2$, i.e. the wheel loads must be applied at equal distances from the middle. We then have

$$m + m' = \frac{P}{\pi + \dfrac{2a}{1-a}} \qquad (1.432)$$

A single load is more dangerous when $2a/(1-a) > \pi$, i.e. $a/l > \pi/(2 + \pi) = 0{\cdot}61$. For simply supported slabs, we put $m' = 0$.

The yield pattern used is not correct because the equilibrium equation for the quadrangular slab part is not satisfied as the moment about a is not zero. It will be seen that we need some upward-directed nodal forces at the transition to the curved yield line. A correct pattern can be obtained by the use of logarithmic spirals (YLT, page 107).

As we will limit ourselves to the symmetrical position, the yield pattern must have yield lines parallel with a. This results in the pattern shown in Fig. 1.434.

As the spiral has to touch both the edges and the yield line perpendicular to these, the two radius vectors become perpendicular to each other, and

$$r_2 = r_1 e^{(\pi/2) \cot \alpha}$$

$$r_1 \sin \alpha = \tfrac{1}{2}(1 - a), \; h = r_2 \sin \alpha$$

$$y = \tfrac{1}{2}(1 - a) + r_2 \cos \alpha$$

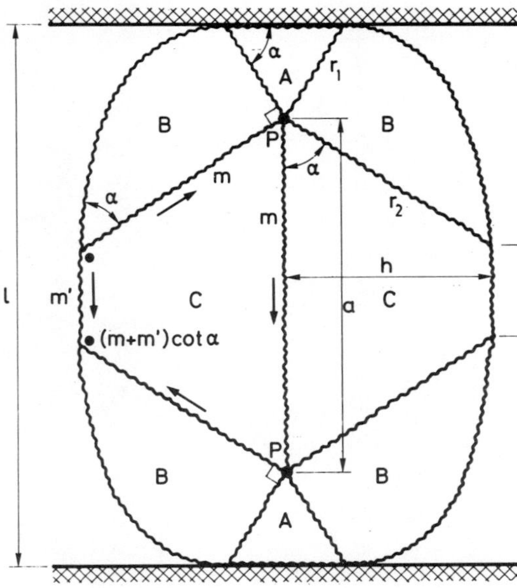

Fig. 1.434

The wheel loads are distributed with P_A, P_B and P_C on the slab parts shown. The moment about the edge now gives

$$P_A = 2(m + m') \cot \alpha$$

Further, formula (79c) in YLT, page 107, gives

$$P_B = \frac{m + m'}{\sin^2 \alpha} \frac{\pi}{2}$$

The equation of projection for C yields

$$P_C = (m + m') \cot \alpha$$

whilst from the moment equation about a, we get

$$(m + m')(1 - 2y) = 2(m + m') \cot \alpha \, h$$

When y and h are expressed by l and a, we obtain the following equation for determination of $\cot \alpha$:

$$\frac{a}{1-a} = 2 \cot \alpha \, e^{(\pi/2) \cot \alpha} \qquad (1.433)$$

after which P is found by means of $P = P_A + 2P_B + 2P_C$, i.e.

$$P = 4(m + m')\left[\cot \alpha + \frac{\pi}{4}(1 + \cot^2 \alpha)\right] \qquad (1.434)$$

When $P = 2\pi(m + m')$, one load is more dangerous. This gives $\cot \alpha = 0{\cdot}55$; $a/l = 0{\cdot}72$.

With $\cot \alpha$ as parameter, we now get related values of a/l and $P/(m + m')$:

cot α	$\frac{a}{l}$	$\frac{P}{m+m'}$ (exact)	$\frac{P}{m+m'}$ (approx.)
0	0	3·14	3·14
0·1	0·19	3·57	3·61
0·2	0·35	4·07	4·23
0·3	0·49	4·62	5·06
0·4	0·60	5·24	6·14
0·5	0·69	5·93	6·28
0·55	0·72	6·28	

These figures show that the approximate formula (1.432) results in rather serious errors at the higher values of a/l. This can be remedied by means of the following correction:

$$m + m' = \frac{P}{\pi + \frac{1{\cdot}4a}{l - a}} \qquad (1.435)$$

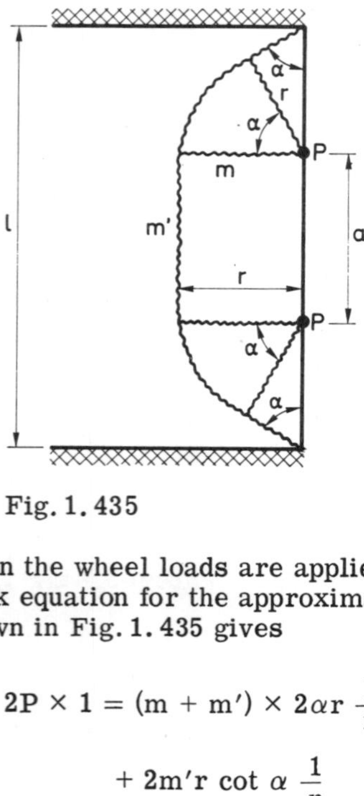

Fig. 1.435

When the wheel loads are applied on the edge, the work equation for the approximate yield pattern shown in Fig. 1.435 gives

$$2P \times 1 = (m + m') \times 2\alpha r \frac{1}{r} + m'a \frac{1}{r} + 2m'r \cot\alpha \frac{1}{r}$$

where $r = \frac{1}{2}(l - a)\sin\alpha$. We then get

$$P = (m + m')\alpha + m' \cot\alpha + m' \frac{a}{l - a} \frac{1}{\sin\alpha}$$

$dP/d\alpha = 0$ gives the following for determination of α:

$$\cos^2\alpha + \frac{m'}{m + m'} \frac{a}{l - a} \cos\alpha = \frac{m}{m + m'} \qquad (1.436)$$

The most economical reinforcement is obtained at $dP/dm' = 0$, as $d(m + m') = 0$. Thence we get $\cos\alpha + a/(l - a) = 0$, i.e. $\alpha > \pi/2$, which is impossible. If (1.436) is used, $\alpha = \pi/2$ gives m = 0 and

$$m' = \frac{2P}{\pi + \frac{2a}{l - a}}$$

i.e. twice the value obtained from (1.432). When the correct yield pattern with the spirals is used, we can therefore reckon with twice the value of (1.435), i.e.

$$m = 0, m' = \frac{2P}{\pi + \frac{1{\cdot}4a}{l - a}} \qquad (1.437)$$

When the loads are distributed, we proceed as in 1.42.

It might be imagined that the yield pattern in Fig. 1.434, turned 90°, would be more dangerous than the pattern in Fig. 1.431. However, it will be found that $\cot\alpha = 2a/3l$ in such a case, and (1.434) then gives smaller moments than (1.431).

1.5 Cantilever slabs (restrained along one side)

1.51 Line load

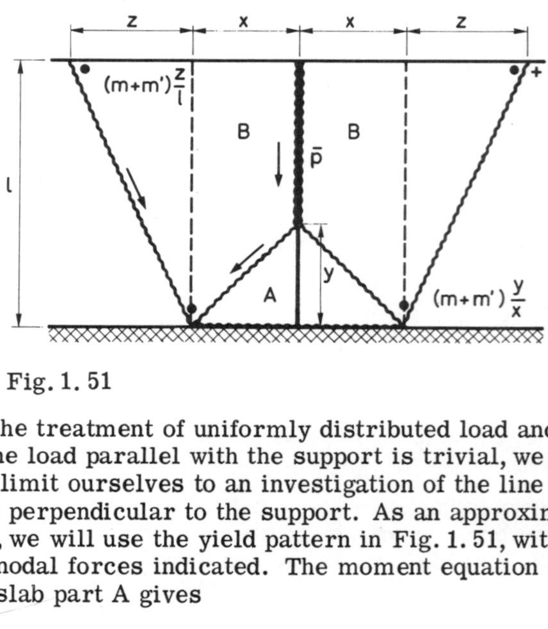

Fig. 1.51

As the treatment of uniformly distributed load and a line load parallel with the support is trivial, we can limit ourselves to an investigation of the line load perpendicular to the support. As an approximation, we will use the yield pattern in Fig. 1.51, with the nodal forces indicated. The moment equation for slab part A gives

$$(m + m') \times 2x = \tfrac{1}{2}\bar{p}y^2 \qquad (1)$$

The equation of projection for B gives

$$(m + m')\left(\frac{z}{l} + \frac{x}{y}\right) = \frac{1}{2}\bar{p}(l - y) \qquad (2)$$

The moment equation for B about $\bar{p}$ gives

$$(m + m')\, l = (m + m')\frac{z}{l}(x + z) + (m + m')\frac{x}{y}x \qquad (3)$$

The moment about the support for A and both B's gives

$$\frac{1}{2}\bar{p}l^2 = 2m'(x+z) + 2(m+m')\frac{z}{l}l \tag{4}$$

For the sake of simplicity we will here introduce the following quantities:

$$\xi = \frac{x}{l}, \eta = \frac{y}{l}, \zeta = \frac{z}{l}, m + m' = \frac{1}{4}\bar{p}lk, \lambda = \frac{m'}{m+m'} \tag{5}$$

and we then get

$$k\xi = \eta^2, 1 - \eta = \frac{1}{2}k\left(\zeta + \frac{\xi}{\eta}\right), 1 = \zeta(\xi + \zeta) + \frac{\xi}{\eta}\xi,$$

$$1 = k[\zeta + \lambda(\xi + \zeta)] \tag{1'-4'}$$

Using the first equation, we can eliminate ξ, thus getting from the second:

$$\eta = \frac{2}{3}(1 - \frac{1}{2}k\zeta)$$

after which, the last gives

$$(k\zeta)^2\lambda + (9 + 5\lambda)k\zeta + 4\lambda - 9 = 0 \tag{1.511}$$

from which $k\zeta$ can be determined and thus also η and ξ. From (3′) we now get

$$k^2 - k^2\zeta^2 = \eta^2(\eta + k\zeta) = \frac{8}{27}(1 + k\zeta)\left(1 - \frac{1}{2}k\zeta\right)^2$$

For determination of k and thus also of $m + m'$, we find from this that

$$k^2 = k^2\zeta^2 + \frac{8}{27}\left(1 + k\zeta\right)\left(1 - \frac{1}{2}k\zeta\right)^2 \tag{1.512}$$

from which we get the following values:

$\frac{m}{m'}$	$k\zeta$	k	$\frac{\bar{p}l}{m+m'}$	$\frac{\bar{p}l}{m'}$
0	0·348	0·628	6·36	6·36
1/4	0·467	0·688	5·81	7·26
1/2	0·525	0·722	5·54	8·30
3/4	0·552	0·739	5·41	9·46
1	0·593	0·765	5·23	10·46

As k is proportional to the amount of steel required, it will be seen that the most economical reinforcement is obtained for $m = 0$. As the values are approximate, we may use the slightly conservative values given by

$$\frac{\bar{p}l}{m+m'} \sim 6 - \frac{m}{m'} \tag{1.513}$$

We see that m' in a restrained slab with $m = 0$ is the same as m in a simply supported slab with double the span and $m' = 0$, (1.411), i.e. exactly the same as in the case of the beam with the same loading.

1.52 Point loads

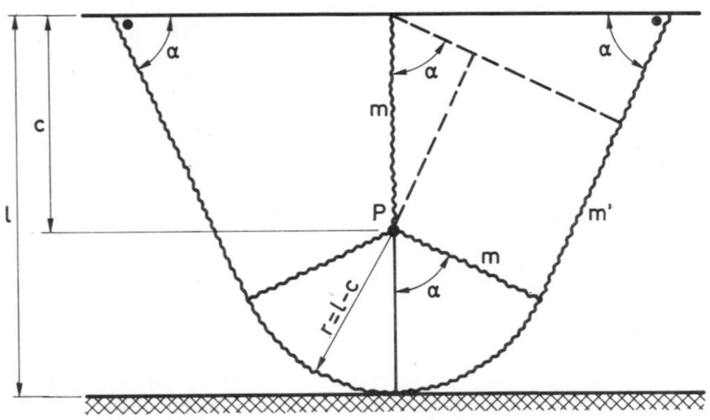

Fig. 1.521

When a point load P is applied at a distance c from the free edge, the work equation for the yield pattern shown (corresponding to Fig. 1.421) gives

$$P \times 1 = 2(m + m')\alpha + 2(m + m')c \sin\alpha \frac{1}{1-c}$$

$$+ 2m'(1 - c + c\cos\alpha)\cot\alpha\frac{1}{1-c}$$

since the rotation is $1/(1-c)$. From this we get

$$\frac{P}{2(m+m')} = \alpha + \frac{c}{1-c}\sin\alpha$$

$$+ \frac{m'}{m'+m}\left(1 + \frac{c}{1-c}\cos\alpha\right)\cot\alpha \tag{1.521}$$

Putting $dP/d\alpha = 0$, the maximum principle now gives

$$\left(1 + \frac{c}{1-c}\cos\alpha\right)\sin^2\alpha$$

$$= \frac{m'}{m'+m}\left[1 + \frac{c}{1-c}\cos\alpha\,(1 + \sin^2\alpha)\right] \tag{1.522}$$

which results in a cubic equation for $\cos\alpha$. However, for $m = 0$, we get as a special case, $\alpha = \pi/2$. The most economical reinforcement can be found by differentiating (1.521) with respect to $m'/(m'+m)$, and we then again get $\alpha = \pi/2$.

Thence,

$$m = 0, m' = \frac{P}{\pi + \frac{2c}{1-c}}; \frac{c}{l} \leqslant \frac{\pi}{2+\pi} = 0{\cdot}61 \tag{1.523}$$

When $2c/(1-c)$ exceeds π, we must use $m' = P/2\pi$.

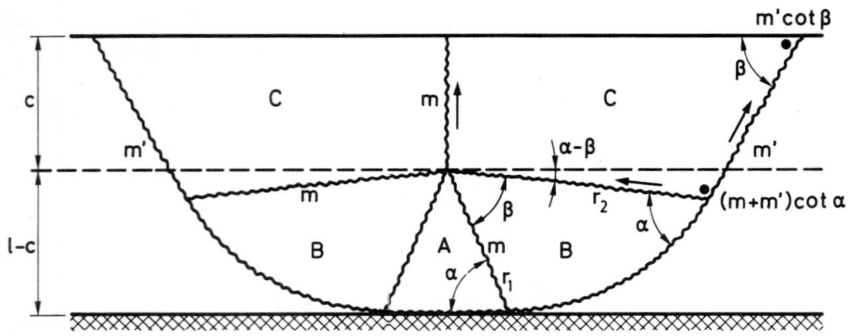

Fig. 1. 522

It will now be seen that we have a formula corresponding to (1. 432). It is also obvious that, if we divide into two halves the slab that is restrained on two sides and subjected to two equal wheel loads applied symmetrically, we get the same moment as in the cantilevered slab with a single point load. This will be realized by considerations analogous to those made in connexion with Fig. 1. 212.

As in 1. 43, the yield pattern is not correct, but the pattern shown in Fig. 1. 522, with logarithmic spiral at B, is correct. P is divided between the slab parts with P_A, P_B and P_C. As in 1. 43, we get

$$P_A = 2(m + m') \cot \alpha, P_B = \frac{m + m'}{\sin^2 \alpha} \beta$$

Then, from the projection for slab part C, we get

$$P_C = (m + m') \cot \alpha + m' \cot \beta$$

The moment equations for C about the axes through P, parallel and perpendicular to the edge, respectively, now give

$$m r_2 \cos (\alpha - \beta) - m'[r_2 \sin (\alpha - \beta) + c] \cot \beta$$
$$+ (m + m') \cot \alpha \, r_2 \sin (\alpha - \beta)$$
$$- m' \cot \beta \times c = 0$$

and

$$(m + m')[r_2 \sin (\alpha - \beta) + c]$$
$$= (m + m') \cot \alpha \, r_2 \cos (\alpha - \beta)$$
$$+ m' \cot \beta \left(\frac{r_2}{\sin \beta} \sin \alpha + c \cot \beta\right)$$

where

$$r_1 \sin \alpha = 1 - c, r_2 = r_1 e^{\beta \cot \alpha}$$

From these, α and β can be determined, after which $P = P_A + 2P_B + 2P_C$ gives the moment. We will now limit ourselves to the economically reinforced slab with m = 0. The moment equations thereby give

$$\frac{c \sin \alpha}{r_2} = -\frac{\sin^2 (\alpha - \beta)}{2 \cos \beta}$$

$$= \frac{\sin^2 \beta \cos (2\alpha - \beta) - \cos \beta \sin^2 \alpha}{\sin^2 \beta - \cos^2 \beta}$$

$$= \frac{c}{1 - c} \sin^2 \alpha \, e^{-\beta \cot \alpha}$$

As the expressions have to be positive, $\cos \beta$ must be negative, i.e. $\beta > \pi/2$. Thus we do not obtain the solution corresponding to Fig. 1. 434. With β as parameter, the equations give the ratio between c and m'. Thus, from the purely trigonometrical equation, we get

$$\cot \alpha = \cot \beta \text{ and } \cot \alpha + 3 \cot \beta = 0$$

the first of which cannot be used as $\alpha < \pi/2$, but $\beta > \pi/2$. With the last, we get

$$\frac{c}{1 - c} = -\frac{8 \cot \beta}{\sqrt{1 + \cot^2 \beta}} e^{-3\beta \cot \beta}$$

$$\frac{P}{m'} = 2\beta(1 + 9 \cot^2 \beta) - 10 \cot \beta$$

which gives

$\cot \beta$	$\frac{c}{l}$	$\frac{P}{m'}$ (exact)	$\frac{P}{m'}$ (approx.)
0	0	π	π
—0·05	0·334	3·61	4·14
—0·10	0·554	4·20	5·62
—0·15	0·691	4·92	2π
—0·20	0·780	5·74	2π
—0·25	0·834	6·65	2π

These values show that the approximation (1. 523) is rather poor. It can be improved by replacing the factor 2 by $\pi/4$, whereby

$$m' = \frac{P}{\pi\left(1 + \frac{c}{4(1 - c)}\right)} \text{ for } \frac{c}{l} \leq 0{\cdot}8 \qquad (1.524)$$

Otherwise, $m' = P/(2\pi)$.

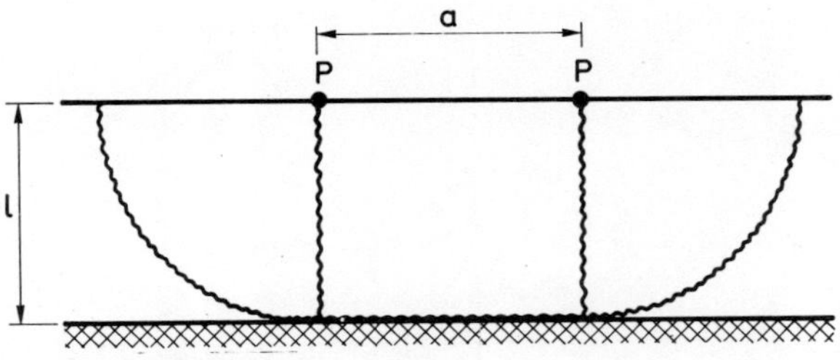

Fig. 1.523

When the loads are distributed, we proceed as in section 1.42. The treatment of wheel loads becomes analagous to that in section 1.43. Thus, for two equal wheel loads on the free edge, with the yield pattern shown in Fig. 1.523, where m = 0, we get

$$2P \times 1 = 2m' \frac{\pi}{2} 1 \frac{1}{1} + m' a \frac{1}{1}$$

$$m' = \frac{2P}{\pi + \frac{a}{1}} \qquad (1.525)$$

When a > (πl), the case of one load becomes decisive, and $m' = P/\pi$.

1.6 Supported on one side and columns

As these slabs are mainly encountered in the form of roof structures, where the dead weight is the dominant load, they will only be calculated for the case of a uniformly distributed load.

1.61 Column at the middle

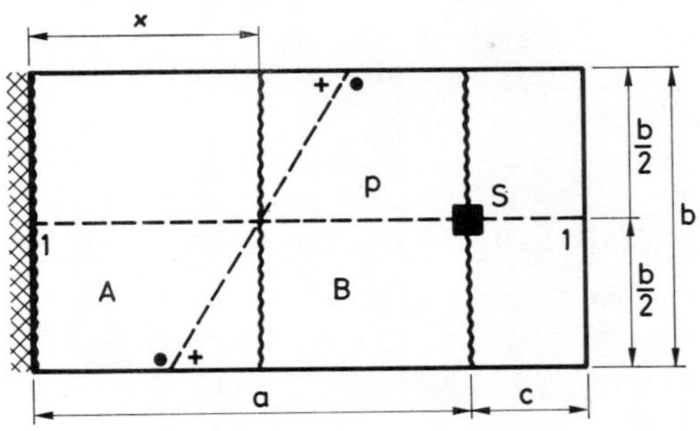

Fig. 1.611

By taking the moment about the centre-line 1-1 through the column for slab part B, it will be seen that equilibrium is not possible for an oblique yield line as the moment from the yield line acts in the same direction as the load.

For the yield line parallel to b we get, as in the case of a beam,

$$m(1 + i) = \frac{1}{2} px^2, \quad m = \frac{1}{2} p \left[(a - x)^2 - c^2\right] \qquad (1.611)$$

Introducing the reduced span l by means of

$$m = \frac{1}{8} pl^2 \qquad (1.612)$$

we get $l\sqrt{1 + i} = 2x$ from the first of the equations in (1.611). Then, eliminating m from both of them, we find

$$l = \frac{2(a^2 - c^2)}{a\sqrt{1 + i} + \sqrt{a^2 + ic^2}} \qquad (1.613)$$

whereby m is again determined by means of (1.612).

The column reaction is statically determinate. We find

$$S = \frac{\frac{1}{2} pb(a + c)^2 - imb}{a} \qquad (1.614)$$

With the yield pattern in Fig. 1.611, the negative moment at the cantilever end will be

$$m' = \frac{1}{2} pc^2 \qquad (1.615)$$

By using the formula in example 12 in YLT, page 32, we get

$$m' = \frac{1}{6} p\left(\frac{b^2}{4} + c^2\right) \qquad (1.616)$$

which gives a higher value when $c < b/(2\sqrt{2}) = 0{\cdot}35b$.

When $a < b^2/4c$, the yield line will intersect the support, and (1.616) will no longer be applicable. With a yield line that just intersects the edge we get $m' = \frac{1}{6}ph^2$, where $h = (a + c) \sin v$ and $\tan v = b/2a$, i.e.

$$m' = \frac{1}{6} p\, (a + c)^2 \frac{b^2}{b^2 + 4a^2} \qquad (1.617)$$

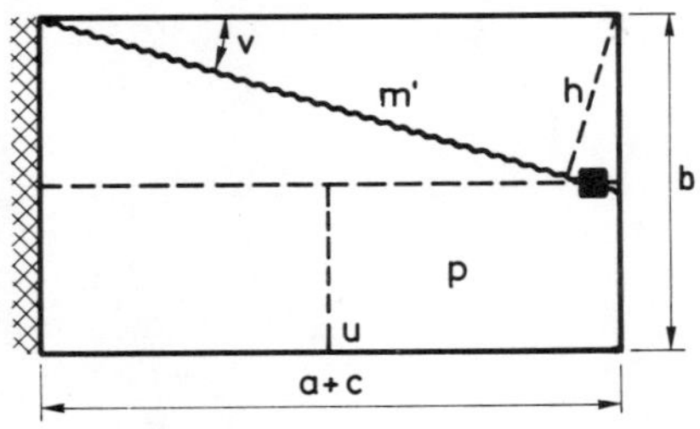

Fig. 1.612

If the column is located near the edge, i.e. if c is small, the local conditions at the column must be investigated by means of formula (90), page 123 in YLT.

We can easily find the deflexions at the centre-line 1-1 on the basis of the positive moment m. The deflexion increment when we move from the centre to the edge is found by means of the negative moment m', as the strip parallel with b can be considered as geometrically restrained at the centre-line because of the symmetry.

We thereby get (cf. Fig. 1.612)

$$EIu \sim \frac{1}{8} ma^2 + \frac{1}{3} m' \frac{b^2}{4} \qquad (1.618)$$

the negative moments being assumed to increase linearly from zero at the edge. For these roof structures, the requirements to stiffness can be somewhat relaxed in relation to the requirements in 1.17; the permissible value may, for instance, be doubled.

Numerical example

A simply supported roof slab measuring 4×6 m, with the column 2 m from the edge, whereby $a = b = 4$ m and $c = 2$ m, is estimated to be 10 cm thick. The total load, including snow and wind, then amounts to 400 kg/m^2. With $i = 0$, we get $l = 3$, $m = \frac{1}{8} \times 400 \times 3^2 = 450$ kg, whilst $m' = \frac{1}{2} \times 400 \times 2^2 = 800$ kg, which, with the usual stresses, gives a slab thickness of 10 cm. The deflexion is therefore

$$u = \frac{\frac{1}{8} \times 450 \times 400^2 + \frac{1}{3} \times 800 \times 2^2}{2 \times 10^5 \times \frac{1}{12} \times 10^3}$$

$$= 1{\cdot}2 \text{ cm} < \frac{400}{250} = 1{\cdot}6 \text{ cm}$$

1.62 Column at the corner

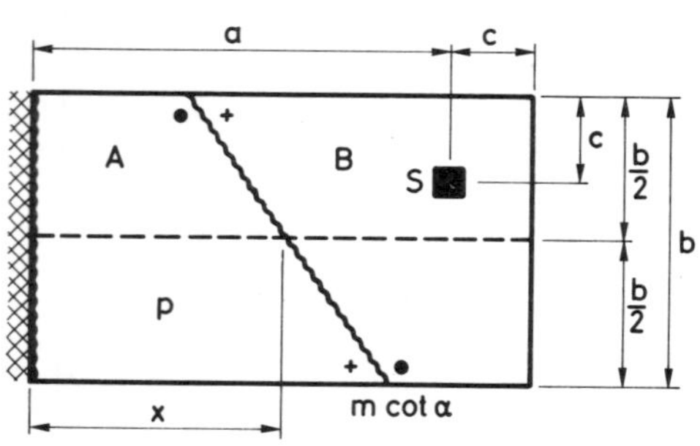

Fig. 1.621

Let us assume that the column is located at the same distance c from both the free edges, and that a is so large in relation to b that the yield pattern shown is formed. We then get the moment equation for slab part A about b:

$$m(1 + i)\, b - m \cot \alpha\, b \cot \alpha$$

$$= \frac{1}{2} pbx^2 + 2 \times \frac{1}{2} p \frac{b}{2} \frac{b}{2} \cot \alpha \frac{b}{6} \cot \alpha$$

or

$$m(1 + i - \cot^2 \alpha) = \frac{1}{2} px^2 + \frac{p}{24} b^2 \cot^2 \alpha \qquad (1)$$

For slab part B, we get the following moment equation about an axis parallel with b through the column:

$$mb - m \cot \alpha\, b \cot \alpha$$

$$= \frac{1}{2} pb(a - x)^2 + \frac{pb^3}{24} \cot^2 \alpha - \frac{1}{2} pbc^2$$

or

$$m(1 - \cot^2 \alpha)$$

$$= \frac{1}{2} p \left(a - x\right)^2 - \frac{1}{2} pc^2 + \frac{pb^2}{24} \cot^2 \alpha \qquad (2)$$

For the same slab section, taking moments about the centre-line parallel with a, we obtain

$$mb \cot \alpha + m \cot \alpha \times b$$

$$= \left(\frac{b}{2} - c\right) S - 2 \frac{p}{2} \frac{b}{2} \frac{b}{2} \cot \alpha \times \frac{b}{3}$$

or, as S is still given by (1.614),

$$m\left(2 \cot \alpha + i \frac{b - 2c}{2a}\right) = \frac{p}{4} \frac{(a + c)^2}{a} (b - 2c)$$

$$- \frac{pb^2}{12} \cot \alpha \qquad (3)$$

Subtracting (2) from (1), we get

$$im = \frac{1}{2} p \left(2ax - a^2 + c^2\right) \qquad (4)$$

which, when inserted in (3), gives

$$m = \frac{p}{4} (b - 2c)(a + c - x) \tan \alpha - \frac{pa^2}{24} \qquad (5)$$

For $i = 0$, we get, in particular, from (4), $x = (a^2 - c^2)/2a$, which is the same value as when the column is located at the centre. With this value we arrive, after some calculation, at the following solution:

$$\tan \alpha = K + \sqrt{K^2 + 1}$$

$$K = \frac{3(a^2 - c^2)^2 + a^2b^2}{6a(a + c)^2(b - 2c)}$$

$$m = \frac{1}{8} pl^2 \left(1 + \frac{a(b - 2c)}{(a - c)^2} \cot \alpha\right) \qquad (1.621)$$

where l is the reduced span determined by means of 1.613, with $i = 0$. We now find the following approximate formula by means of comparative calculations:

$$m = \frac{1}{8} pl^2 \frac{a - 1{\cdot}6c}{a - 0{\cdot}35b - 0{\cdot}9c} \quad (a \geqslant b) \qquad (1.622)$$

For $c = \frac{1}{2}b$, this passes into (1.612), which applies for the column at the centre.

When $i > 0$, we may naturally expect (1.622) to retain its validity. This is obviously the case for $c = \frac{1}{2}b$, so it must be sufficient to investigate the conditions for the other limiting case, $c = 0$. With this value and $i = 1$, we get, in particular, with $x = k(a/2)$

$$\tan\alpha = \frac{b}{a}\,\frac{2-k}{8(k-1)-k^2}$$

$$\frac{b^2}{a^2} = 3(k-1)\,\frac{8(k-1)-k^2}{k^2-5(k-1)}$$

$$m = \frac{1}{8}\,pl^2(1+\sqrt{2})^2\,(k-1) \qquad (1.623)$$

With k as parameter, we get from this related values of m and b/a. Comparison with (1.622) shows that the above assumption is correct, i.e. (1.622) also applies to restrained slabs.

The formula is only derived for $a \geqslant b$, as we would otherwise provide two columns.

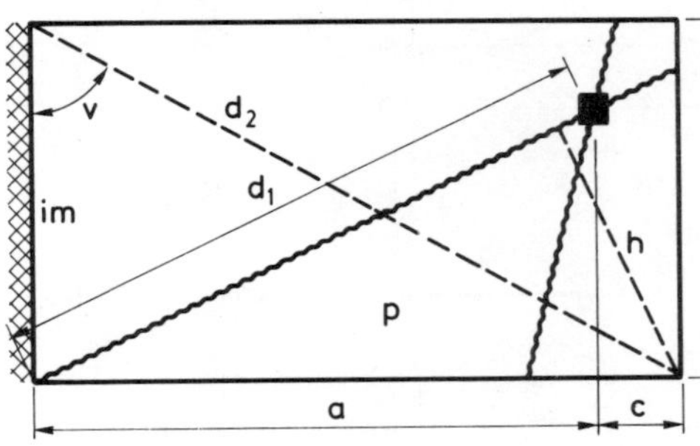

Fig. 1.622

By means of the yield pattern in Fig. 1.622, with the yield line going from the corner of the slab to the column, we find the negative moment

$$m' = \frac{1}{6}\,ph^2 = \frac{1}{6}\,p\,\frac{(a+c)^2\,(b-c)^2}{a^2+(b-c)^2} \qquad (1.624)$$

When $a > (b-c)^2/c$, we get the formula corresponding to (1.616):

$$m' = \frac{1}{6}\,p\,[(b-c)^2+c^2] \qquad (1.625)$$

When the column is located near the centre-line, a more dangerous yield line, going from the long side of the slab through the column may form, as shown in Fig. 1.622. A detailed investigation of this is rather complicated, so we will confine ourselves to just giving the result. The investigation proves that we can, to a good approximation, assume that

$$m' = \frac{1}{4}\,pbc,\; c > \frac{1}{3}\,b \qquad (1.626)$$

The local conditions at the column are dealt with in equation (90) in YLT. The column reaction is determined as mentioned, still using (1.614).

The maximum deflexion normally occurs at the free corner that is furthest from the column. One strip is not sufficient to determine this deflexion, but if we consider the two strips in Fig. 1.622, d_1 through the column, and the diagonal d_2 to the corner at which the deflexion is to be determined, we find that d_1 is approximately perpendicular to the positive yield line, and d_2 to the negative yield line. At the point of intersection between the two strips, we have the deflexion u_1, which is determined on the basis of the positive moment by means of $EIu_1 \sim 1/8\; md_1^2$. Fig. 1.623 shows the strip d_2 and its

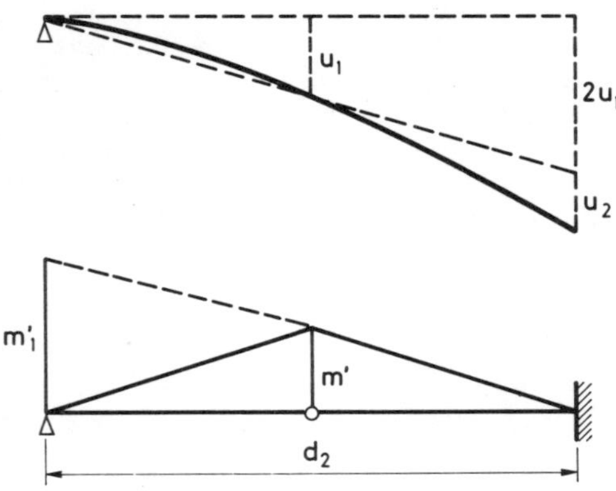

Fig. 1.623

conjugate beam, which serves to determine u_2. When the slab is simply supported, we get the following fictitious support moment for the triangular moment surface shown, which forms the fictitious load:

$$EIu_2 = \frac{1}{3}\,m'\left(\frac{d_2}{2}\right)^2 + \frac{1}{3}\,m'\,\frac{d_2}{2}\,\frac{d_2}{2} = \frac{1}{6}\,m'd_2^2$$

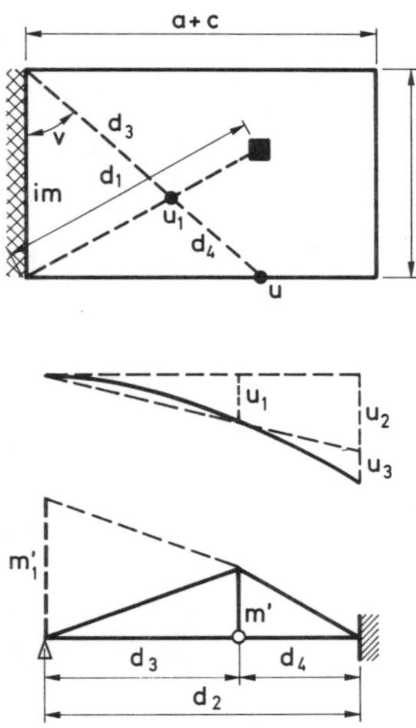

Fig. 1.624

If the slab is restrained corresponding to m'_1, we further get the following increment due to the dotted curve:

$$\frac{1}{6}\,m'_1\,\frac{d_2}{2}\,\frac{d_2}{2} = \frac{1}{24}\,m'_1 d_2^2 = \frac{1}{24}\,imb^2$$

since $m'_1 = im\cos^2 v$, $\cos v = b/d_2$. The total deflexion is $u = 2u_1 + u_2$, which gives

$$EIu \sim \frac{1}{4}\, m[a^2 + (b - c)^2] + \frac{1}{6}\, m'[(a + c)^2 + b^2]$$

$$+ \frac{1}{24}\, imb^2 \qquad (1.627)$$

This formula assumes the column to be located near the corner. When the column is closer to the centre, it is no longer certain that the maximum deflexion is found at the corner. Fig. 1. 624 shows which strips are to be used. It also shows the strip $d_2 = d_3 + d_4$ and its conjugate beam. In the same way as above, with m'_1 at the support, we get

$$EIu_1 \sim \frac{1}{8} md_1{}^2 = \frac{1}{8} m\, [a^2 + (b - c)^2]$$

$$u_2 = u_1 \frac{d_2}{d_3}$$

$$EIu_3 = \frac{1}{6} m'_1 d_3 d_4 + \frac{1}{3} m' d_3 d_4 + \frac{1}{3} m' d_4{}^2$$

When u_1 is near the centre of d_1, we have

$$d_2{}^2 \sim b^2 \left(1 + \frac{a^2}{(b + c)^2}\right), \frac{d_3}{d_2} \sim \frac{b + c}{2b}$$

$$\frac{d_4}{d_2} \sim \frac{b - c}{2b}$$

As $m'_1 = im \cos^2 v$, we get

$$EIu \sim \frac{1}{4}\, m \frac{b}{b + c} [a^2 + (b - c)^2]$$

$$+ \frac{1}{6}\, m'b\, (b - c) \left[1 + \frac{a^2}{(b + c)^2}\right] + \frac{1}{24}\, im\, (b^2 - c^2)$$

$$(1.628)$$

Numerical example

We take the same roof as in 1. 61, but with the column at the corner, 1 m from both edges. We thus get a = 5m and l = 4·8 m, after which (1. 622) gives $m = \frac{1}{8} \times 400 \times 4{\cdot}8^2 \times (5 - 1{\cdot}6)/(5 - 1{\cdot}4 - 0{\cdot}9)$ = 1450 kg. Accordingly, h = 14 cm, so we must recalculate with the load 500 kg/m² instead of 400 kg/m². We then get $m = \frac{5}{4} \times 1450 = 1810$ kg. By means of (1. 625) we find that $m' = \frac{1}{6} \times 500 \times (3^2 + 1^2) = 833$ kg. In accordance with (1. 614), the column load will be 7200 kg. Neglecting the capital, equation (90a) in YLT, with $\omega = \pi/2$, gives

$$S = 1{\cdot}57 \times 1810 + 0{\cdot}43 \times 833$$

$$+ 4 \times 1 \times \sqrt{500 \times 2643} = 7800 \text{ kg} > 7200 \text{ kg}$$

i.e. the local yield pattern at the column is not critical.

We find the deflexion by means of (1. 627):

$$u = \frac{\frac{1}{4} \times 1810\, (300^2 + 500^2) + \frac{1}{6} \times 833\, (600^2 + 400^2)}{2 \times 10^5 \times \frac{1}{12} \times 14^3}$$

$$= 5{\cdot}0 \text{ cm} > \frac{500}{250} = 2{\cdot}0 \text{ cm}$$

Despite the considerable increase in the slab thickness, we get far too great a deflexion.

When c is small, we can with good approximation disregard the slab parts shown in Fig. 1. 625 and calculate the slab as though c = 0, but with b' instead of b. In this way, with b' = b − c, we find the approximate expression:

$$m = \frac{1}{8}\, pl^2 \frac{a}{a - 0{\cdot}35\, (b - c)} \qquad (1.629)$$

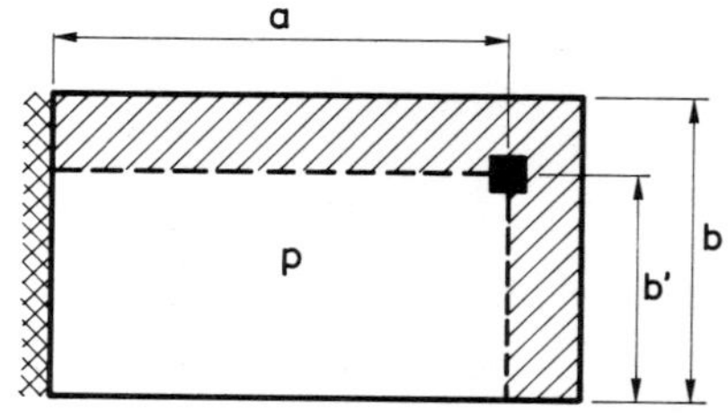

Fig. 1. 625

where $l = 2a/(1 + \sqrt{i + 1})$. We thus find a somewhat higher value of m. However, as the above calculation does not take into account the live load, m will actually be somewhat higher, so (1. 629) can be used in cases in which the live load has to be considered. The difference between the two values does not exceed 10% when $c < 0{\cdot}45b - 0{\cdot}2a$.

1.63 Columns at both corners

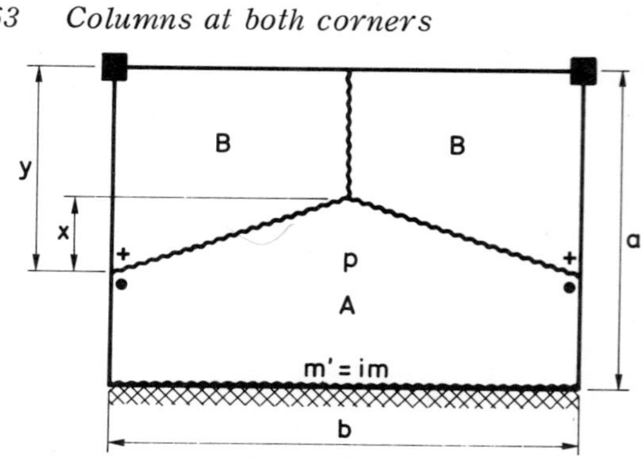

Fig. 1. 631

As it must be possible to consider this case in the same way as above, we will at first limit ourselves to considering the case where the columns are located right at the corners. With the yield pattern shown in Fig. 1. 631, where the nodal forces are 2m(x/b), a moment equation about the support yields for slab part A:

$$m(1 + i)\, b + 2 \times 2m \frac{x}{b} (a - y)$$

$$= \frac{1}{2}\, pb(a - y)^2 + \frac{1}{2}\, pbx \left(a - y + \frac{x}{3}\right)$$

or

$$m\left(1 + i + 4\frac{x(a-y)}{b^2}\right)$$

$$= \frac{1}{2}p\left[(a-y)^2 + x\left(a - y + \frac{x}{3}\right)\right] \qquad (1)$$

For slab part B, a moment equation about edge b gives

$$mb - 2 \times 2m\frac{x}{b}y = \frac{1}{2}pby^2 - \frac{1}{2}pbx\left(y - \frac{x}{3}\right)$$

or

$$m\left[1 - 4\frac{xy}{b^2}\right] = \frac{1}{2}p\left[y^2 - x\left(y - \frac{x}{3}\right)\right] \qquad (2)$$

Subtracting (2) from (1), we get

$$m\left(i + 4\frac{xa}{b^2}\right) = \frac{1}{2}pa(a - 2y + x) \qquad (3)$$

Finally, a moment equation about a for the same slab part yields

$$my = \frac{1}{8}pyb^2 - \frac{1}{2}p\frac{b}{2}x\frac{b}{3}$$

or

$$my = \frac{pb^2}{24}(3y - 2x) \qquad (4)$$

Dividing (3) by (4), we get

$$\frac{\frac{ib^2}{4a} + x}{y} = \frac{a - 2y + x}{y - \frac{2}{3}x} = \frac{2y - a - \frac{ib^2}{4a}}{\frac{2}{3}x}$$

or

$$2y^2 - \left(a - \frac{ib^2}{4a}\right)y = \frac{2}{3}x^2 + \frac{1}{6}xi\frac{b^2}{a} \qquad (5)$$

Dividing (2) by (4), we get

$$\frac{\frac{b^2}{4} - xy}{y} = \frac{y^2 - x\left(y - \frac{x}{3}\right)}{y - \frac{2}{3}x} = \frac{\frac{b^2}{4} - y^2 - \frac{x^2}{3}}{\frac{2}{3}x}$$

or

$$y\left(\frac{b^2}{4} - y^2\right) = \frac{1}{6}b^2x - \frac{1}{3}yx^2 \qquad (6)$$

We now eliminate x^2 from (5) and (6), and find

$$x = \frac{3}{2}y\,\frac{2 - \frac{y}{a}\left(4\frac{a^2}{b^2} - i\right)}{2 + \frac{y}{a}i} \qquad (7)$$

which, inserted in (6), gives the following equation:

$$\left[i^2 - \frac{3}{4}\left(4\frac{a^2}{b^2} - i\right)^2\right]\left(\frac{y}{a}\right)^3 + \left(i + 12\frac{a^2}{b^2}\right)\left(\frac{y}{a}\right)^2$$

$$+ (1 - i)\frac{y}{a} = 2 \qquad (8)$$

from which y can be determined, and then x by means of (7) and m by means of (4).

If the slab is simply supported, we get the special case,

$$\left(\frac{y}{a}\right)^3 - \frac{b^2}{a^2}\left(\frac{y}{a}\right)^2 - \frac{1}{12}\frac{b^4}{a^4}\frac{y}{a} + \frac{1}{6}\frac{b^4}{a^4} = 0 \qquad (9)$$

We can avoid having to solve a cubic equation by adopting y/a as parameter. We then get a quadratic equation for b^2/a^2.

Related values of m and b/a can be calculated by means of equations (8) and (9). These values show that we have, to a good approximation,

$$m = \frac{pab}{18 + 12i}\left(1 + \frac{5 + 2i}{4}\frac{a}{b}\right),\ \frac{b}{a} < 2 - \frac{1}{2}i$$

$$\text{for } i \leqslant 1 \qquad (1.631)$$

When x = 0, we get the yield pattern in Fig. 1.611. We then find from equations (5) and (6) that

$$y = \frac{a}{2} - \frac{ib^2}{8a} = \frac{b}{2},\ b = \frac{2a}{1 + \sqrt{1 + i}} = 1$$

which is equation (1.613) with c = 0. From (4) we find $m = \frac{1}{8}pb^2 = \frac{1}{8}pl^2$. When $b < 1$, we get $m = \frac{1}{8}pl^2$.

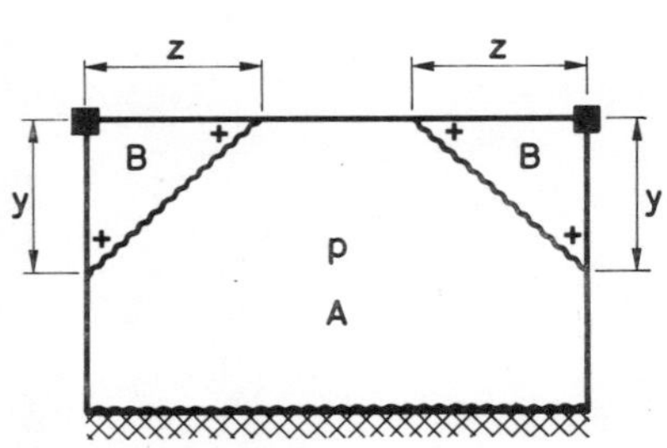

Fig. 1.632

For the yield pattern in Fig. 1.632, the moment equations for B about the free edges show that these can only be satisfied when y = z = 0. As the column reaction is

$$S = \frac{\frac{1}{2}pba^2 - imb}{2a} \qquad (1.632)$$

we get $m = m' = \frac{1}{2}S$ (YLT, page 85, at top), i.e.

$$m = \frac{pab}{8 + 2i\frac{a}{b}} \qquad (1.633)$$

The transition from (1. 631) to (1. 633) takes place around b = a. Thus, the yield pattern shown in Fig. 1. 631 is only likely to arise in a few cases. However, as we can prevent local failure at the column by providing this with a suitable capital, we can use (1. 631) for higher values of b. If the capital found necessary is too large, a capital of suitable size can be selected, but then, of course, the local conditions at the column will be more critical. These conditions are investigated by means of equation (90) in YLT, which thus replaces (1. 633). However, the yield pattern in Fig. 1. 631 assumes that $x < y$, which gives the limit for b/a indicated in (1. 631).

When the slab is simply supported, the yield patterns in Figs. 1. 633 and 1. 634 may occur if the corners of the slab are not anchored or if no top reinforcement is provided.

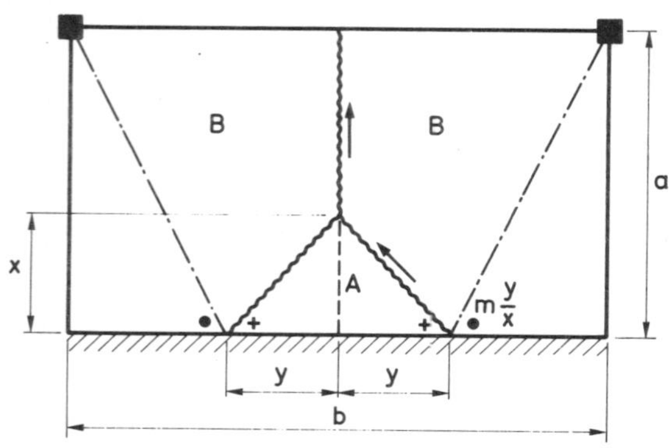

Fig. 1. 633

In the first case, we get at slab part A:

$$m = \frac{1}{6} px^2 \tag{1}$$

For slab part B, a moment equation about the free edge gives

$$my + m\frac{y}{x} a = \frac{1}{2} p \frac{b}{2} a^2 - \frac{1}{2} pxy \left(a - \frac{x}{3}\right)$$

Inserting m from (1) in this equation, we get

$$xy = \frac{3}{8} ab \tag{2}$$

For the same slab part, a moment equation about the free edge a gives

$$ma + m\frac{y}{x}\left(\frac{b}{2} - y\right) = \frac{1}{2} pa \frac{b^2}{4} - \frac{1}{2} pxy \left(\frac{b}{2} - \frac{y}{3}\right)$$

Inserting in this equation m and y from (1) and (2), we get

$$x^3 = \frac{9}{32} ab^2 \tag{3}$$

which again gives

$$m = \frac{pab}{16} \sqrt[3]{\frac{3b}{2a}} \tag{1.634}$$

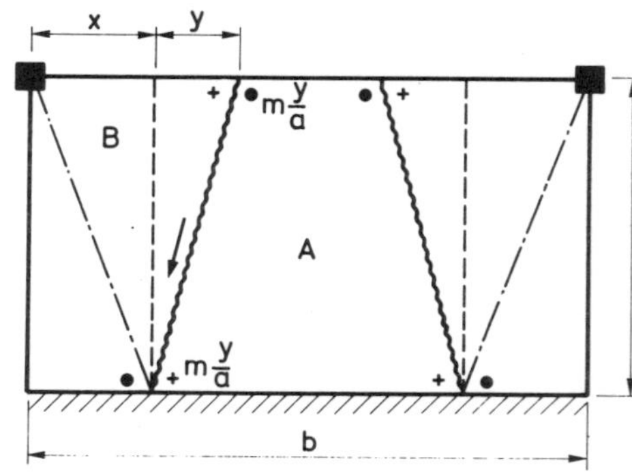

Fig. 1. 634

The other yield pattern has already been dealt with in section 1. 21, and (1. 217) gives m, but only for load on one side of the axis of rotation, whereas we have all the time assumed load over the entire slab. Continuing with the latter assumption, the equations on page 22 will be altered as follows:

$$2my = \frac{1}{2} pxa^2 + \frac{1}{6} pya^2 \tag{B1}$$

$$ma - m\frac{y^2}{a} = \frac{1}{6} pa \left[x^2 + x(x + y) + (x + y)^2\right] \tag{B2}$$

$$4my = \frac{1}{2} pa^2(b - 2x) - \frac{1}{3} pa^2\, 2y \tag{A}$$

Eliminating x and y and with $k = pa^2/m$, we get

$$k^2 + 2\left(3 - 4\frac{a^2}{b^2}\right)k - 192\frac{a^2}{b^2} = 0 \tag{1.635}$$

from which k and thus m can be determined.

Now, comparing (1. 631), (1. 634) and (1. 635), we find that (1. 634) will never be decisive and (1. 635) only when $b > 1{\cdot}8a$, but in this case, the slab will have such a shape that it would be most natural to provide more than two columns.

The yield patterns shown in Fig. 1. 623 and 1. 624 might also arise in the case of a restrained slab, since there would be a negative yield line at the axis of rotation. In the latter Figure, m can immediately be calculated by means of (1. 218), which, however, gives a lower value of m than (1. 631).

In slabs encountered in practice, we can thus disregard the yield patterns in Figs. 1. 633 and 1. 634.

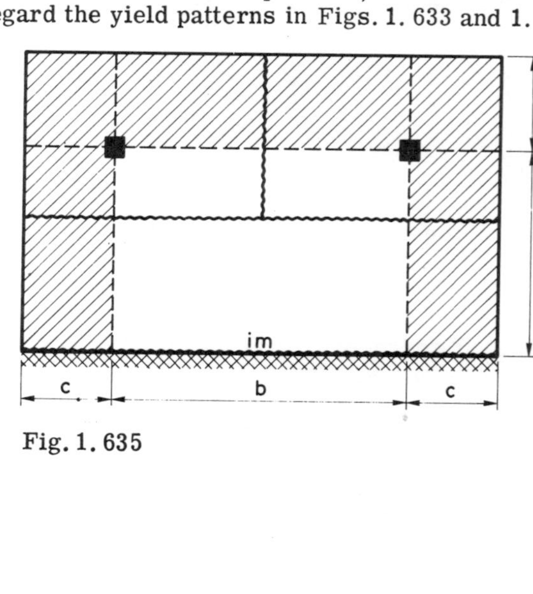

Fig. 1. 635

When the columns are located at some distance from the corners, the calculations become rather difficult, and we shall therefore not reproduce them here. It will be on the safe side to neglect the hatched part of the slab, i.e. we can use (1. 631), as might be expected in accordance with the result at the end of section 1. 62. However, a comparison with the exact calculation shows that this is only a good approximation when the columns are located rather close to the corners. At greater distances, we get too conservative a solution. It is found that we can find m by interpolating between the value m_0 for $c = 0$ and the value of m_1 that is found when c is so large that we get the yield pattern shown in Fig. 1. 635.

This is precisely the pattern that forms the transition to the cases in which the slab is simply calculated as a beam that spans from the support to the columns, i.e. by means of formulae (1. 612) and (1. 613). In the transitional case, we obviously have

$$m = \frac{1}{8}pl^2 = \frac{1}{8}pb^2 - \frac{1}{2}pc_1^2$$

i.e.

$$l = \frac{2(a^2 - c_1^2)}{a\sqrt{1+i} + \sqrt{a^2 + ic_1^2}} = \sqrt{b^2 - 4c_1^2} \qquad (1)$$

In the special case of simple support, we get

$$c_1^2 = a(b - a) \qquad (2)$$

$$m_1 = \frac{1}{8}p(2a - b)^2 \qquad (3)$$

This solution is normally only practicable when the slab is almost square, but it can be used for the purpose of interpolation. As we know that (1. 631) can be utilized for small values of c, linear interpolation cannot be used. With quadratic interpolation we get

$$m = m_0 - (m_0 - m_1)\left(\frac{c}{c_1}\right)^2 \qquad (4)$$

where m_0 is the value corresponding to $c = 0$ that is given by (1. 631). Inserting this and m_1 and c_1 from (2) and (3), we get

$$m = \frac{p}{72}\left[4ab + 5a^2 - 31c^2 + 9\frac{b}{a}c^2\right], i = 0 \qquad (1.636)$$

When $i > 0$, the exact solution of (1) cannot be found when c_1 is considered as unknown. On the other hand, b/a can be expressed by means of c_1, and if we do this we find that quadratic interpolation between the value corresponding to $c_1 = 0$ and that corresponding to $c_1 = a$ is very accurate. As the former value is $2/(1 + \sqrt{1+i})$ and the latter is 2, we thus have, to a good approximation,

$$\frac{b}{a} = \frac{2}{1 + \sqrt{1+i}} + \frac{c_1^2}{a^2}\left(2 - \frac{2}{1 + \sqrt{1+i}}\right)$$

or

$$\frac{c_1^2}{a^2} = \frac{\frac{b}{a} - \frac{2}{1+\sqrt{1+i}}}{2 - \frac{2}{1+\sqrt{1+i}}} \sim \frac{\frac{b}{a} - 1 + \frac{i}{6}}{1 + \frac{i}{6}}, i \leq 1 \qquad (2a)$$

and

$$m_1 = \frac{1}{8}p\left[b^2 - 4a^2\,\frac{\frac{b}{a} - 1 + \frac{i}{6}}{1 + \frac{i}{6}}\right], i \leq 1 \qquad (3a)$$

With these values for c_1 and m_1, and m_0 in accordance with (1. 631), we can find m by means of (4). For $i = 1$, we get

$$m = \frac{p}{120}\left[4a^2 + 7ab - 50c^2 + 17{\cdot}5\frac{b}{a}c^2\right], i = 1 \qquad (1.637)$$

If we now compare the values found by means of (1. 636) and (1. 637) with the exact values, we will find that the approximation is on the safe side. In slabs likely to arise in practice, the deviation does not exceed 10%. In the least favourable case (with the columns near the centre), the deviation does not, moreover, exceed about 20%.

As above, the local conditions at the column must be investigated by means of equation (90) in YLT. The negative moment is $m' = \frac{1}{2}pc^2$.

The deflexions are estimated in the following manner. At a strip between the columns, the maximum deflexion between these is

$$EIu_1 \sim \frac{1}{8}mb^2$$

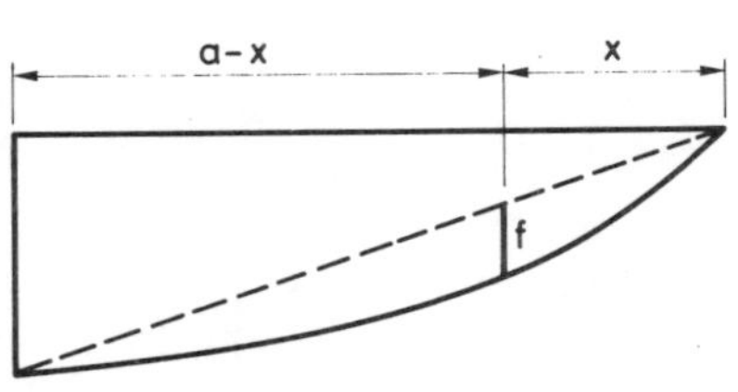

Fig. 1. 636

For the strip half-way between the columns and parallel with a, we have (Fig. 1. 636),

$$EIf \sim \frac{1}{2}mx(a - x)$$

We then find the maximum total deflexion u from

$$u = u_1\frac{x}{a} + f$$

We thereby find that

$$EIu \sim \frac{1}{8}m\left(a + \frac{b^2}{4a}\right)^2, (b \leq 2a) \qquad (1.638)$$

A diagonal strip gives, more simply but somewhat less accurately,

$$EIu \sim \frac{1}{8}\, m(a^2 + b^2)$$

Numerical example

According to (1.636), a simply supported roof slab, with c = 1 m, a = 3 m, b = 4 m and p = 400 kg/m², will get the following moment:

$$m = \frac{400}{72}\left[4 \times 3 \times 4 + 5 \times 3^2 - 31 \times 1^2 + 9 \times \frac{4}{3} \times 1^2\right] = 411 \text{ kg}$$

In example 21 in YLT, the exact value is found to be 400 × 0·96 = 384 kg. The column investigation is given on page 132 in YLT. With a slab thickness of 10 cm, we find the following deflexion by means of (1.638):

$$u = \frac{411\left(300 + \dfrac{400 \times 400}{4 \times 300}\right)^2}{8 \times 2 \times 10^5 \times \dfrac{1}{12} \times 10^3}$$

$$= 2{\cdot}4 \text{ cm} > \frac{400}{250} = 1{\cdot}6 \text{ cm}$$

i.e. the thickness should be increased to 12 cm.

1.7 Supported on two adjacent sides and a column

1.71 Column at the free corner

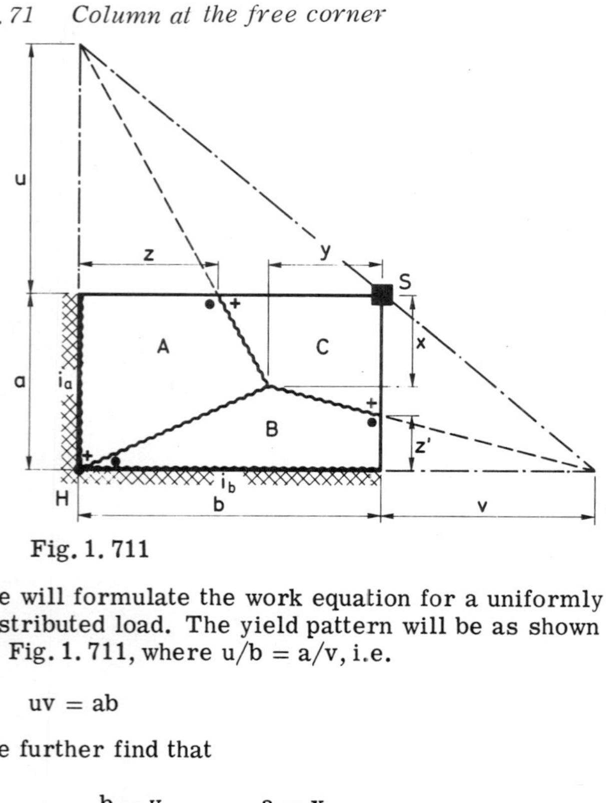

Fig. 1.711

We will formulate the work equation for a uniformly distributed load. The yield pattern will be as shown in Fig. 1.711, where u/b = a/v, i.e.

$$uv = ab$$

We further find that

$$z = u\frac{b - y}{u + x},\; z' = v\frac{a - x}{v + y}$$

Unit deflexion at the point of intersection (x, y) will result in the following rotations of slab parts A, B and C:

$$\theta_A = \frac{1}{b - y},\; \theta_B = \frac{1}{a - x},$$

$$\theta_{C,a} = \frac{1}{y + x\dfrac{v}{a}},\; \theta_{C,b} = \frac{1}{x + y\dfrac{u}{b}}$$

the last two of which are components of θ_C with regard to a and b. The internal moments are

$$M_A^i = -m(1 + i_a)a,\; M_B^i = -m(1 + i_b)b$$

$$M_{C,a}^i = -m(a - z'),\; M_{C,b}^i = -m(b - z)$$

The external moments are

$$M_A^u = \frac{1}{6}\, p(a + u)(b - y)^2 - \frac{1}{6}\, puz^3$$

$$M_B^u = \frac{1}{6}\, p(b + v)(a - x)^2 - \frac{1}{6}\, pv(z')^3$$

$$M_{C,a}^u = \frac{1}{6}\, p(a - x - z')y^2 + \frac{1}{6}\, px[y^2 + y(b - z) + (b - z)^2]$$

$$M_{C,b}^u = \frac{1}{6}\, p(b - y - z)x^2 + \frac{1}{6}\, py[x^2 + x(a - z') + (a - z')^2]$$

Formula (37), YLT, page 60, is used for the last two.

After rearrangement, the work equation gives

$$m = \frac{pab}{6}\,\frac{2 + \dfrac{x}{a}\dfrac{u}{u + x} + \dfrac{y}{b}\dfrac{v}{v + y}}{\dfrac{a(1 + i_a)}{b - y} + \dfrac{a}{v + y} + \dfrac{b(1 + i_b)}{a - x} + \dfrac{b}{u + x}} \qquad (1.711)$$

We will first use this to find approximate values of m with the aid of estimated yield patterns. If the axis of rotation is assumed to lie at an angle of 45° with the sides, we get u = b and v = a. We will further assume that (x, y) lies at an equal distance from the supported sides, i.e. that a − x = b − y = k(a + b), where k is an unknown constant that is determined by means of the maximum principle. With these values, (1.711) gives the following moment for simple support:

$$m = \frac{pab}{6}\, k\left[3 - \left(2 + \frac{a}{b} + \frac{b}{a}\right)k\right]$$

which will be maximum for k = 3/[2(2 + a/b + b/a)].

We thereby obtain

$$m_o = \frac{3pab}{8\left(2 + \frac{a}{b} + \frac{b}{a}\right)}, i_a = i_b = 0 \quad (1.712)$$

Assuming the yield line to lie along one diagonal and the axis of rotation to lie parallel with the other, we get $u = a, v = b$ and $x/a = y/b = k$. If we are using (1.711), with $i = 0$, this gives

$$m = \frac{P}{6\left(\frac{a}{b} + \frac{b}{a}\right)}(1-k)(1+2k)$$

$$m_{max} = \frac{3P}{16\left(\frac{a}{b} + \frac{b}{a}\right)}, k = \frac{1}{4}$$

Taking the axis of rotation as perpendicular to the diagonal to the column, and the yield line from the corner as lying symmetrical to the same diagonal with regard to the bisector, we get $u = b^2/a, v = a^2/b$ and $(a - x)/b = (b - y)/a = k$ and

$$m = \frac{P}{12} k\left[3 - k\left(\frac{a}{b} + \frac{b}{a}\right)\right]$$

with the same maximum as above when $k = 3/[2(a/b + b/a)]$.

We have obtained the same value of m for two different yield patterns. Therefore, a pattern lying between the two must be expected to give a better approximation, and it is just such a pattern that led to (1.712), which must thus provide a particularly good approximation. It will easily be seen that this gives a higher value of m than the other formula.

In the case of restrained slabs, the same yield pattern gives

$$m = \frac{pab}{6} \frac{k\left[3 - \left(2 + \frac{a}{b} + \frac{b}{a}\right)k\right]}{1 + (1-k)\frac{ai_a + bi_b}{a+b}}$$

where the exact determination of k leads to a quadratic equation.

By substituting in the numerator the same value of k as for simple support (a value that deviates only slightly from 3/8 as long as a and b do not differ greatly), and in the denominator $1 - k = 5/8 \sim 0{\cdot}6$, we get

$$m = \frac{m_o}{1 + 0{\cdot}6\frac{ai_a + bi_b}{a+b}} \quad (1.713)$$

where m_o is determined by means of (1.712). The error in these formulae is only a few per cent, as will be seen from the exact calculation in the following.

When the point of intersection (x, y) in Fig. 1.711 falls outside the slab, we get the yield pattern in Fig. 1.712. Precisely as in the corresponding yield pattern, Fig. 1.632, the yield line at C must be immediately adjacent to the column, and $S = 2m$. The moment equation for A is then

$$m(1 + i_a)a - m\frac{x}{a}x = \frac{1}{6}pax^2$$

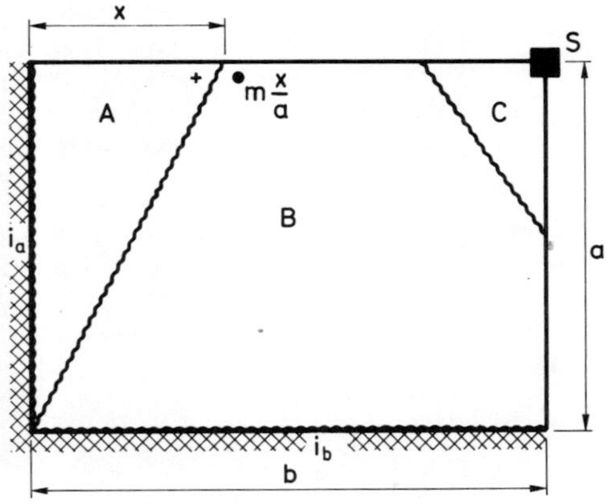

Fig. 1.712

For B, we get the moment equation

$$mi_bb + mx + m\frac{x}{a}a = \frac{1}{2}pa^2b - \frac{1}{3}pa^2x - Sa$$

These are solved by

$$\frac{x}{b} = \frac{3}{1 + \sqrt{1 + K}}, K = 3\frac{3 + 2\frac{a}{b} + i_b}{(1 + i_a)\frac{a^2}{b^2}}$$

$$m = \frac{1}{6}pa^2\frac{3b - 2x}{2(x + a) + i_bb}$$

For simple support, we get the special case

$$m_o = \frac{pab}{8\left(1 + \frac{a}{3b}\right)}, a \leqslant b \quad (1.714)$$

Equation (1.713) can be used to a good approximation for restrained slabs.

A comparison between (1.712) and (1.714) shows that they both give the same value of m for a = b, but apart from this, (1.714) gives the higher value of m. The yield pattern in Fig. 1.712 is therefore the more dangerous and, as (1.713) applies for both yield patterns, this is also the case for restrained slabs. However, the possibility of local failure at the corner can be avoided by providing the column with a suitable capital, and then equation (90) in YLT, page 123, must be used instead of (1.714). We can thus as a rule use (1.712) to determine m and (90) to determine the capital.

A more accurate determination of the yield pattern

is necessary for calculating the column reaction S. The unknown quantities x, y, u and v can be determined by means of the maximum principle, bearing in mind that u and v are related by uv = ab, i.e. udv + vdu = 0. According to equation (40) in YLT, page 64, we now get

$$\frac{T}{N} = \frac{dT}{dN} = \frac{\frac{\partial T}{\partial u}du + \frac{\partial T}{\partial v}dv}{\frac{\partial N}{\partial u}du + \frac{\partial N}{\partial v}dv} = \frac{u\frac{\partial T}{\partial u} - v\frac{\partial T}{\partial v}}{u\frac{\partial N}{\partial u} - v\frac{\partial T}{\partial v}} \qquad (1)$$

as du/u = — dv/v, in accordance with the above. From (1.711) we get the following new expressions for m:

By differentiating with regard to x in the numerator and denominator:

$$m = \frac{pab}{6}\,\frac{\dfrac{u^2}{a(u+x)^2}}{\dfrac{b(1+i_b)}{(a-x)^2} - \dfrac{b}{(u+x)^2}} \qquad (2)$$

By differentiating with regard to y:

$$m = \frac{pab}{6}\,\frac{\dfrac{v^2}{b(v+y)^2}}{\dfrac{a(1+i_a)}{(b-y)^2} - \dfrac{a}{(v+y)^2}} \qquad (3)$$

By differentiating with respect to u and v, (1) yields

$$m = \frac{pab}{6}\,\frac{u\dfrac{x^2}{(u+x)^2} - v\dfrac{y^2}{(v+y)^2}}{-\dfrac{bu}{(u+x)^2} + \dfrac{av}{(v+y)^2}} \qquad (4)$$

In connexion with (1.711), we have here four equations, which, in principle, determine the unknown quantities, but such a determination is not practicable. However, as the approximate formulae provide us with m, we can find x and y from (2) and (3), estimating u and thus also v. The correctness of u can be checked by substituting the calculated values in (4). When the correct values of x, y, u and v have been determined, the equilibrium equation for the slab part C in Fig. 1.711 gives the column reaction:

$$S = \frac{1}{2}p\left[x^2\frac{b-y}{u+x} + 2xy + y^2\frac{a-x}{v+y}\right] + m\frac{b-y}{u+x} + m\frac{a-x}{v+y} \qquad (1.715)$$

where the last two terms are the contributions from the nodal forces shown.

In the same way, we get the reactions on sides a and b:

$$R_a = \frac{1}{2}p(b-y)\left[a + u - \frac{u^2}{u+x}\right] + m(1+i_a)\frac{a-x}{b-y} - m\frac{b-y}{u+x}$$

$$R_b = \frac{1}{2}p(a-x)\left[b + v - \frac{v^2}{v+y}\right] + m(1+i_b)\frac{b-y}{a-x} - m\frac{a-x}{v+y} \qquad (1.716)$$

At the corner, we get the anchorage force:

$$H = m(1+i_a)\frac{a-x}{b-y} + m(1+i_b)\frac{b-y}{a-x} \qquad (1.717)$$

The longest side can be assumed to be less than twice the shortest side, as we would otherwise arrange more columns, so the calculation is only carried out for $a \leqslant b \leqslant 2a$ in the case of simple supports. At restraints, the yield pattern is impossible for b = 2a.

The results, which are given in the following Table, show clearly that S/P, H/P and $(R_a + R_b)/P$ vary only slightly with a/b, i.e. they can be assumed to be equal to the values for the square slab with the same degree of restraint. Further, we know, with sufficient accuracy for practical purposes, that $R_a/R_b = a\sqrt{1+i_a}/(b\sqrt{1+i_b})$, and as the exact values of P/m are specified, the suitability of the approximate formulae established earlier can be checked.

i_a	i_b	$\frac{a}{b}$	$\frac{u}{a}$	$\frac{x}{a}$	$\frac{y}{b}$	$\frac{P}{m}$	$\frac{S}{P}$	$\frac{R_a}{P}$	$\frac{R_b}{P}$	$\frac{H}{P}$
		1	1	0·25	0·25	10·7	0·212	0·487	0·487	0·187
0	0	2/3	2·2	0·115	0·354	11·0	0·217	0·403	0·563	0·183
		1/2	4·36	0·001	0·415	11·5	0·225	0·344	0·608	0·177
1	1	1	1	0·162	0·162	17·1	0·130	0·551	0·551	0·232
		2/3	1·93	0·035	0·257	17·7	0·130	0·450	0·648	0·228
1	0	1	0·9	0·290	0·112	13·6	0·168	0·603	0·439	0·210
		2/3	1·9	0·176	0·230	13·3	0·172	0·511	0·530	0·213

The following formulae apply to a good approximation:

$$S = \frac{S_o}{1 + 0{\cdot}6\,\dfrac{i_a a + i_b b}{a + b}}, \quad R_a = \frac{Pa\sqrt{1 + i_a}}{a\sqrt{1 + i_a} + b\sqrt{1 + i_b}}$$

$$S_o = 0{\cdot}22\,P \qquad R_b = \frac{Pb\sqrt{1 + i_b}}{a\sqrt{1 + i_a} + b\sqrt{1 + i_b}}$$

(1.718)

H can, as usual, be determined by means of (1.119). Because of the approximations, the check $P + H = R_a + R_b + S$ does not agree exactly.

For the simply supported slab, S can be calculated from the theory of elasticity by means of the results in sections 1.32 and 1.33. As the resultant deflexion due to the column reaction S_o and the load p, according to (1.323) and (1.333), is

$$EIu = \frac{1}{8}\,pa^2b^2 - \frac{1}{2}\,S_oab = 0$$

we get $S_o = 0{\cdot}25$ pab, which is independent of a/b and almost equal to S_o in (1.718).

1.72 *Column inside the free corner*

We will here limit ourselves to the square slab since, as in the previous sections, the results can also be assumed to apply to a rectangular slab. For the sake of simplicity we will assume that $i_a = i_b$, so the yield pattern is symmetrical (Fig. 1.721) and dependent only upon the distance from the point of intersection of the yield lines to the sides. We find that

$$z = \frac{a - c}{2a - x}\,x$$

$$y = 2(a - c)\,\frac{a - x}{2a - x}$$

We can then establish the work equation, assuming

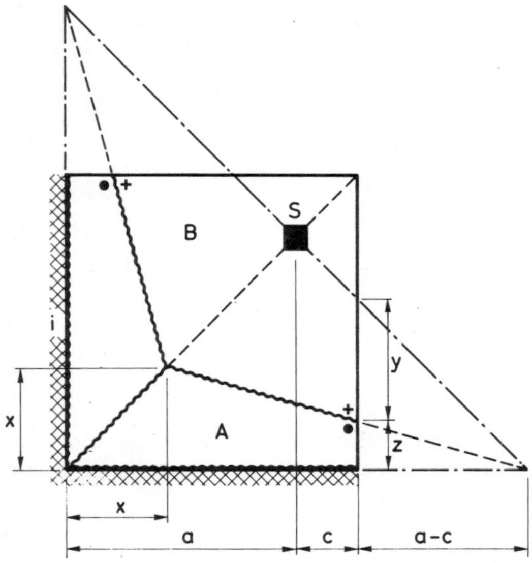

Fig. 1.721

unit deflexion of the point of intersection. This results in the rotations,

$$\theta_A = \frac{1}{x}$$

$$\theta_B = \frac{1}{(a - x)\sqrt{2}}$$

for slab parts A and B. For the internal moments for A and ½B, we get

$$M_i^A = m(1 + i)(a + c), \quad M_i^B = m\left(c\sqrt{2} + y\sqrt{\frac{1}{2}}\right)$$

and for the external moments,

$$M_u^A = \frac{1}{6}\,p \times 2ax^2 - \frac{1}{6}\,p(a - c)z^2$$

$$M_u^B = \frac{1}{6}\,pa\sqrt{2}(a - x)^2 \times 2$$

$$-\frac{1}{6}\,p(a - c)\sqrt{2}\,y^2 \times \frac{1}{2} - \frac{1}{6}\,pc\sqrt{2}\,c^2 \times 2$$

After rearrangement, the work equation gives

$$m = \frac{p}{6}\,x\,\frac{3(a + c)^2(a - c)a - x(a + c)(5a^2 - 2ac - c^2) + 2a^2x^2}{2\,(1 + i)(a + c)a^2 - x(2 + 3i)(a + c)a + [ia + (1 + i)c]x^2}$$

(1.721)

and, differentiating the numerator and denominator with respect to x, we get, by means of the maximum principle,

$$m = \frac{p}{6}\,\frac{3(a + c)^2(a - c)a - 2x(a + c)(5a^2 - 2ac - c^2) + 6a^2x^2}{-(2 + 3i)(a + c)a + 2[ia + (1 + i)c]x}$$

(1.722)

These equations are treated in the same way as in section 1.71. With an estimated value of x, we get an approximate value of m by means of (1.721) that is slightly on the low side. With a slightly higher value of m, (1.722) is used to determine x, which gives a better value of m in (1.721), etc. The calculations have been carried out for i = 0 and i = 1, and the results show that we have, to a good approximation,

$$m = \frac{m_o - 0{\cdot}15pc^2}{1 + 0{\cdot}6\,\dfrac{i_a(a + c) + i_b(b + c)}{a + b + 2c}} \qquad (1.723)$$

where m_o is the value determined by means of (1.712) (using a and b). The effect of the restraint is thus the same as in (1.713) (with $a + c_1$ and $b + c$). The column reaction is found by the equation of projection for slab part B:

$$S = p\left[(a - x)\sqrt{2}\,a\sqrt{2} - (a - c)\sqrt{2}\,y\sqrt{\frac{1}{2}} + 2c^2\right]$$

$$+ 2m\frac{x}{2a - x} \qquad (1.724)$$

The results obtained from the above calculations show that we have, to a good approximation,

$$S = \frac{0{\cdot}22pab}{1 + 0{\cdot}6\,\dfrac{i_a(a + c) + i_b(b - c)}{a + b + 2c}} + \frac{1}{2}p(a + b)c$$

$$+ 2{\cdot}5pc^2 \qquad (1.725)$$

As $H \sim 2m\sqrt{1 + i_a}\,\sqrt{1 + i_b}$, we get $R_a + R_b = p(a + c) \times (b + c) + H - S$, where we can assume that $R_a/R_b = (a + c)\sqrt{1 + i_a}/[(b + c)\sqrt{1 + i_b}]$.

When the column is not equidistant from the edges, but stands at a distance c from edge b and a distance d from edge a, whereby the slab has the side lengths a + c and b + d, the formula must be assumed to be as follows:

$$m = \frac{m_o - 0{\cdot}15\,pcd}{1 + 0{\cdot}6\,\dfrac{i_a(a + c) + i_b(b + d)}{a + b + c + d}} \qquad (1.726)$$

$$S = \frac{0{\cdot}22pab}{1 + 0{\cdot}6\,\dfrac{i_a(a + c) + i_b(b + d)}{a + b + c + d}} + \frac{1}{2}p(ad + bc)$$

$$+ 2{\cdot}5\,pcd \qquad (1.727)$$

This can be seen if we divide the slab as shown in Fig. 1.722. For the innermost part of the slab, with the sides a and b, we get the first term in the expression for S, and for the outer parts, we get the second term pcd. The excess 1·5 pcd can be attributed to the fact that the load acts on the cantilever part of the slab [cf. a simply supported beam with span l and a single cantilever end of length c; the reaction in this beam is then increased by pc + ½p (c²/l)].

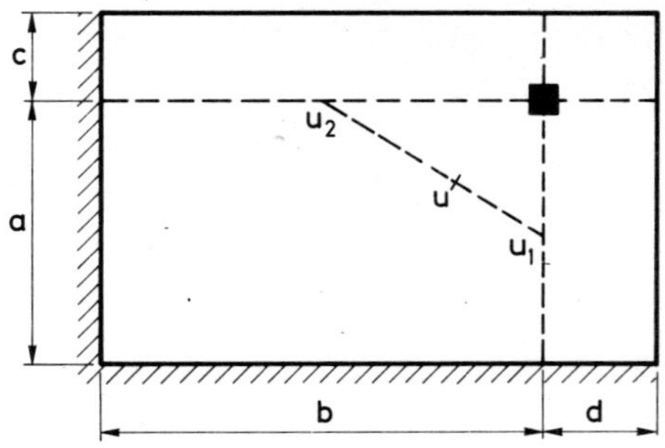

Fig. 1.722

The local conditions at the column must be investigated by means of equation (90) in YLT, page 123. To determine the maximum deflexion, we could simply use a strip along the diagonal to the column, and we would then get $EIu \sim \frac{1}{8}m(a^2 + b^2)$. However, a more accurate value can be obtained by means of the strips shown in Fig. 1.722, which fit the yield pattern better. At the middle of a, we have $EIu_1 \sim \frac{1}{8}ma^2$, and at the middle of b, $EIu_2 \sim \frac{1}{8}mb^2$. The strip from u_1 to u_2 has the dip $EIf \sim \frac{1}{8}m(a^2/4 + b^2/4)$, and as $u \sim f + \frac{1}{2}u_2 + \frac{1}{2}u_1$, we get

$$EIu \sim \frac{3}{32}m(a^2 + b^2) \qquad (1.728)$$

Numerical example

In example 22 in YLT, page 79, a = 4·5 m, b = 3·0 m, c = 1·5 m, and d = 1·0 m. As $i_a = i_b = 0$, we get from (1.726) and (1.727)

$$m = \frac{3p \times 4{\cdot}5 \times 3}{8\left(2 + \dfrac{4{\cdot}5}{3} + \dfrac{3}{4{\cdot}5}\right)} - 0{\cdot}15p \times 1{\cdot}5 \times 1 = 0{\cdot}99p$$

$$S = 0{\cdot}22p \times 4{\cdot}5 \times 3 + \frac{1}{2}p(4{\cdot}5 \times 1 + 3 \times 1{\cdot}5)$$

$$+ 2{\cdot}5p \times 1{\cdot}5 \times 1 = 11{\cdot}2p$$

The exact calculation in YLT gave the same value of m, and S = 11·6p, for the dead load.

With p = 400 kg/m², we get m = 400 kg, which gives h = 8 cm. The deflexion is thus

$$u = \frac{3}{32}\,\frac{400(450^2 + 300^2)}{2 \times 10^5 \times \dfrac{1}{12} \times 8^3} = 1{\cdot}3\text{ cm}$$

$$< \frac{450}{250} = 1{\cdot}8\text{ cm}$$

The slab is thus sufficiently stiff as a roof structure. The investigation of the conditions at the column is given on page 133 in YLT.

1.8 Supported on four columns

With a uniformly distributed load and symmetrically

placed columns, we get one of the yield lines shown in Fig. 1. 81, which gives

$$m_a = \frac{1}{8} p(a^2 - 4c^2)$$

$$m_b = \frac{1}{8} (b^2 - 4d^2) \qquad (1.81)$$

$$S = \frac{P}{4}$$

The negative moments are

$$m_a' = \frac{1}{2} pc^2, m_b' = \frac{1}{2} pd^2 \qquad (1.82)$$

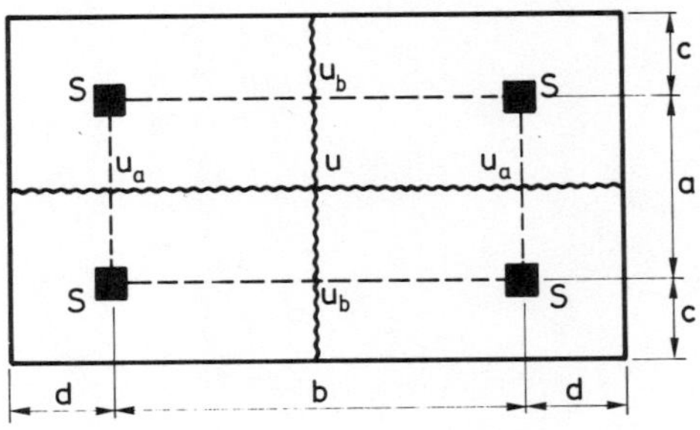

Fig. 1. 81

It will be appropriate to design the slab as an orthotropic slab according to these moments.

The local conditions at the column can be investigated by means of equation (90) in YLT. This only covers isotropic slabs, but by means of an affine transformation, the calculation of the orthotropic slab can be converted to that of an isotropic slab. However, this calls for the same reinforcement ratio at top and bottom, which in turn requires that $a/b = c/d$, which can usually be arranged. The affine slab will then be square, with the largest of the sides as side length.

With $c = 0{\cdot}35a$ and $d = 0{\cdot}35b$, we get, in particular,

$$m_a' = m_a = \frac{1}{16} pa^2, m_b' = m_b = \frac{1}{16} pb^2 \qquad (1.83)$$

At strips between the columns, one parallel with a and the other with b, we get respectively

$$EIu_a \sim \frac{1}{8} m_a a^2, EIu_b \sim \frac{1}{8} m_b b^2$$

The dips for the strip between the u_a's and the strip between the u_b's are $EIf_a \sim 1/8\, m_b b^2$ and $EIf_b \sim 1/8\, m_a a^2$, respectively. As $u = u_a + f_a$ or $u_b + f_b$, it will be seen that both methods yield the same result, i.e.

$$EIu \sim \frac{1}{8}(m_a a^2 + m_b b^2) \qquad (1.84)$$

A somewhat less accurate estimate can again be obtained by utilizing a diagonal strip, which gives $EIu \sim 1/8\, m(a^2 + b^2)$, where m is the greater of the moments m_a and m_b. In the section perpendicular to the diagonal, we have moreover, $m_d = m_a \cos^2 v + m_b \sin^2 v$, and when this is used, we again get (1. 84).

Numerical example

In the case of the square slab corresponding to (1. 83), the total load is $P = p \times 1{\cdot}7^2 a^2 = 2{\cdot}89\ pa^2$, $S = 0{\cdot}722\ pa^2$, $m = m' = pa^2/16 = P/46{\cdot}2$. If a square capital is used, we get $p_c = S/l^2 = 0{\cdot}722\ pa^2/l^2$, and equation (90) gives

$$0{\cdot}722\ pa^2 \left(1 - \sqrt[3]{\frac{l^2}{0{\cdot}722a^2}}\right)$$

$$= 2m + 4 \times 0{\cdot}35a \sqrt{p(m + m')}$$

$$= \frac{1}{8} pa^2 + 1{\cdot}4a \sqrt{\frac{p^2a^2}{8}} = 0{\cdot}625pa^2$$

so that

$$\frac{l}{a} = \frac{(0{\cdot}722 - 0{\cdot}625)^{3/2}}{0{\cdot}722} = \frac{1}{24}$$

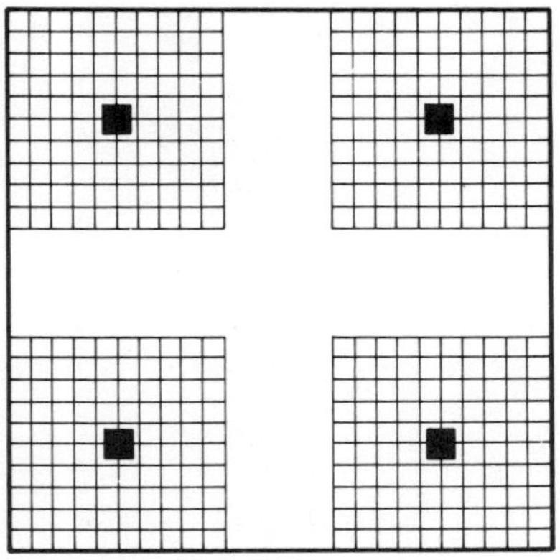

Fig. 1. 82

We must next determine how far the top reinforcement has to be taken past the column because of the local conditions here. As $m_s = m_s' = m$, we can use Table IV on page 135 in YLT for this purpose. As $\xi = 0{\cdot}35a \sqrt{p/m} = 1{\cdot}4$, we get $T = 2{\cdot}9$, where ξ and T are defined by means of equations (86) and (96) in YLT. As $m' = 0$, where the top reinforcement ends, we get $t \sqrt{1/2} = T \sqrt{m/p} = 0{\cdot}72a$. As t is the distance from the corner to the limit of the reinforcement (Fig. 84 in YLT), $t \sqrt{1/2}$ will be its distance from the edge, and $(0{\cdot}72 - 0{\cdot}35)a = 0{\cdot}37a$, the distance that the top reinforcement has to be taken into the inner part of the slab. This is considerably more than the $0{\cdot}15a$ obtained by means of (1. 166) or (0. 16) with $i = 1$. We assume here that the top reinforcement is arranged in square panels at the columns as in the case of flat-slab structures (Fig. 1. 82).

With $p = 400\ kg/m^2$ and $a = 4$ m, we get the moment $m = 400$ kg, which gives a slab thickness $h = 8$ cm.

The deflexion will then be

$$u = \frac{\frac{1}{4} \times 400 \times 400^2}{2 \times 10^5 \times \frac{1}{12} \times 8^3} = 1{\cdot}9 \text{ cm} > \frac{400}{250} = 1{\cdot}6 \text{ cm}$$

The thickness must be increased to 9 cm. The side of the capital $l = 400/24 = 17$ cm. The column reaction is $0{\cdot}72 \times 400 \times 16 = 4600$ kg. The shear stress is $\tau = 4600/(4 \times 17 \times 9) = 7{\cdot}5$ kg/cm^2.

2 Skew slabs

Skew slabs are of minor importance in practice and can usually be dealt with in the same way as rectangular slabs, so the following treatment can be kept quite short.

2.1 Parallelogram-shaped slabs

These are most frequently encountered in skew bridges.

2.11 Supported on all four sides

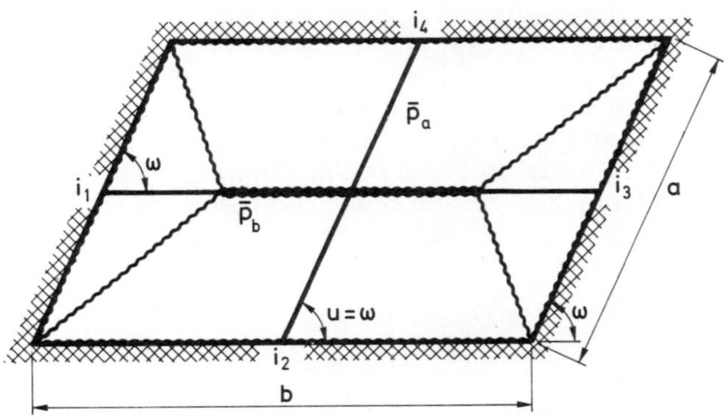

Fig. 2.111

In the case of a uniformly distributed load p and the line loads shown, $\bar{p}_a$ and $\bar{p}_b$, the equilibrium equations for the yield pattern shown in Fig. 2.111 are derived and dealt with in exactly the same way as in section 1.13.

The expressions show that the skew slab has the same moment m as a rectangular slab with the same *spans* and same loads. By span we mean the perpendicular distance between opposite sides, and for the skew slab, we get a sin ω and b sin ω. If $\bar{p}_a$ is not parallel with a, but forms the angle u ≠ ω with b, we get, by means of the corresponding rectangular slab, $\bar{p}_a' = p_a \sin \omega / \sin u$. For p alone, formula (109), page 139 in YLT, gives the same reaction $\bar{q}$ as in the corresponding rectangular slab (with the sides a sin ω, b sin ω). On the other hand, the corner forces are as follows (cf. the derivation in section 2.21):

$$H_{1,2} = \frac{1}{\sin \omega} H'_{1,2} + m(1 + i_1 + 1 + i_2) \cot \omega$$

and correspondingly for $H_{3,4}$,

$$H_{2,3} = \frac{1}{\sin \omega} H'_{2,3} - m(1 + i_2 + 1 + i_3) \cot \omega \tag{2.111}$$

and correspondingly for $H_{1,4}$, where H′ applies to the corresponding rectangular slab.

Multiplication by sin ω of the equation

$$2\bar{q}_a a + 2\bar{q}_b b = pab \sin \omega + \sum H$$

for the skew slab gives the corresponding equation for the rectangular slab, the contributions with cot ω cancelling out.

Because of the acute angle, the influence of the corner levers is somewhat greater than in the case of the rectangular slab. For calculations similar to those in the following chapter 3, we find a correction factor,

$$k = 1{\cdot}75 - \frac{2}{3} \sin \omega \tag{2.112}$$

for ω ⩾ 30°, by which the moment must be multiplied. For example, ω = 60° gives k = 1·17, i.e. an increase of 17%.

In Fig. 2.111 it is assumed that the negative yield lines are located above the supports. However, in certain types of slab, yield lines like AB′ and CD′ in Fig. 2.112 may result in a greater moment because the distance AC′ between these is greater than the distance between the sides AB and CD.

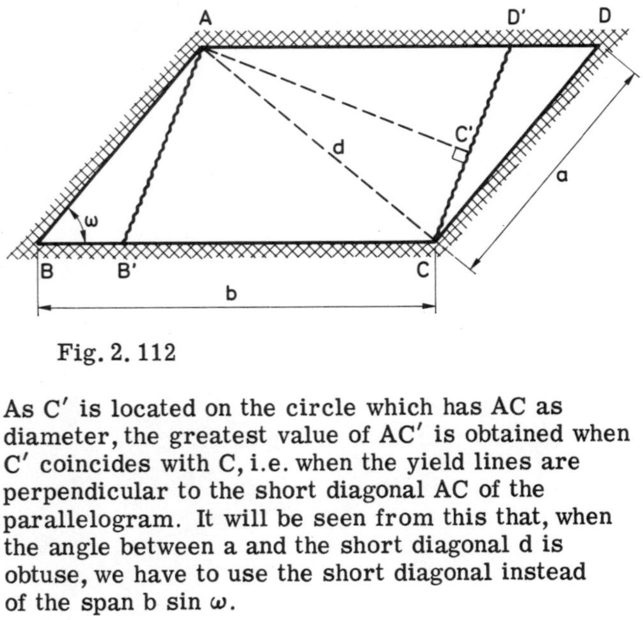

Fig. 2.112

As C′ is located on the circle which has AC as diameter, the greatest value of AC′ is obtained when C′ coincides with C, i.e. when the yield lines are perpendicular to the short diagonal AC of the parallelogram. It will be seen from this that, when the angle between a and the short diagonal d is obtuse, we have to use the short diagonal instead of the span b sin ω.

For a point load, formula (1.14) applies without alteration, as the derivation shows that only the number of restrained sides, and not the shape of the slab, is of importance.

In the main, the interpolation formulae (1.146-8) would apply if the wheel loads were distributed over parallelograms with the sides k and l parallel with a and b (Fig. 2.113). However, no significant change in m can be expected if these parallelograms are replaced by rectangles, the sides of

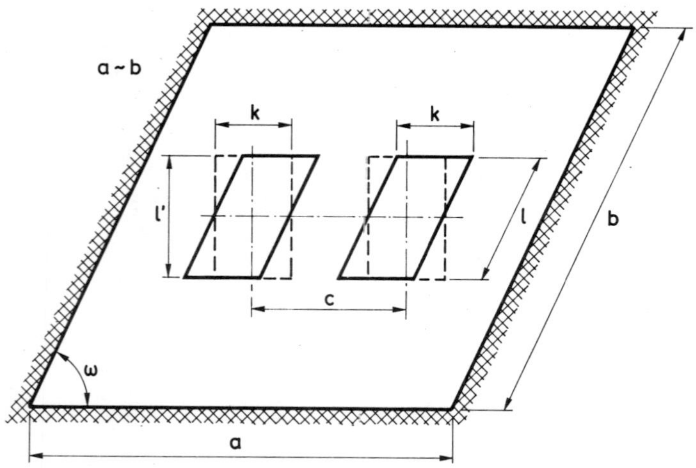

Fig. 2.113

which are $l' = l \sin \omega$ and k, as shown in the Figure. In other words, formulae (1.146-8) are used with $l = l'/\sin \omega$. In the limiting case, $k = l' = 0$, formula (1.146) gives $m = P/6$, which differs less than 5% from the correct value, $P/2\pi$. In the limiting case, $k = l = a = b$, we get $m = P/24$, whilst the correct value is

$$m = k\frac{pa^2 \sin^2 \omega}{24} = \frac{P}{24} \sin \omega \left(1{\cdot}75 - \frac{2}{3} \sin \omega\right)$$
$$= \frac{P}{24}\,(1 \pm 0{\cdot}08)$$

for $\omega > 45°$, in which the corner levers are taken into account.

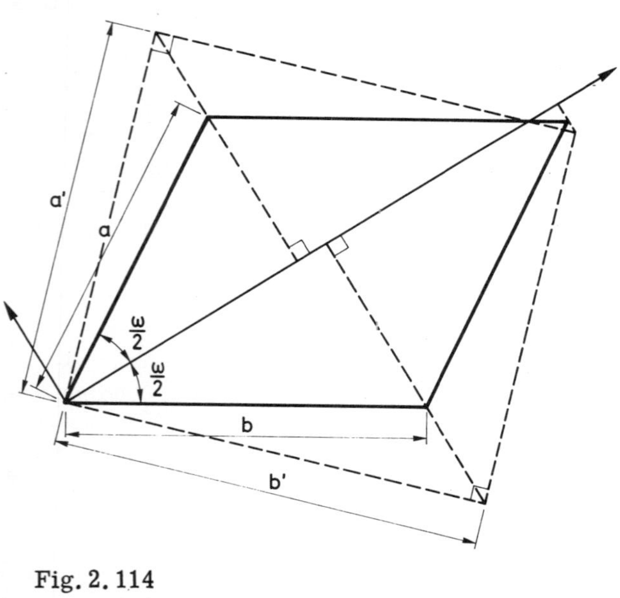

Fig. 2.114

It would be natural to arrange the reinforcement parallel with the sides, but this would be very uneconomical. If the same reinforcement, corresponding to m_ω, were provided in the direction of each side, then in accordance with YLT, page 56, the slab would have to be treated as an orthotropic slab with reinforcement parallel with and perpendicular to the bisector, in the ratio $\tan^2 (\omega/2)$. In accordance with the affinity theorem, m_ω could then be calculated by means of an isotropic slab affine to the given slab in the ratio $\cot (\omega/2)$. This slab (see Fig. 2.114) is rectangular, with the sides $a' = a\sqrt{2} \cos (\omega/2)$ and $b' = b\sqrt{2} \cos (\omega/2)$, which have the same ratio as a and b and as the spans $a \sin \omega$ and $b \sin \omega$. The moment m_ω and the moment m that is found in the skew slab when this is isotropically reinforced would then have the same ratio as the loads on the rectangular slabs $a'b'$ and $a \sin \omega \times b \sin \omega$, i.e.

$$\frac{m_\omega}{m} = \frac{2 \cos^2 \frac{\omega}{2}}{\sin^2 \omega} = \frac{1}{1 - \cos \omega} \qquad (2.113)$$

It will be seen that $\omega = 60°$ would already give $m_\omega = 2m$; and even so slight a degree of skewness as corresponding to $\omega = 80°$ would result in a 20% increase in the reinforcement if we were to reinforce in the direction of the sides.

2.12 Supported on two sides

For a uniformly distributed load p, we get the following moment for the yield pattern shown in Fig. 2.121:

$$m = \frac{1}{8} p l_r^2 \qquad (2.121)$$

where the reduced span l_r is determined from l by means of formula (0.11). If b forms an obtuse angle with the short diagonal, we must use d instead of l, as the negative yield lines are not formed along the supports but perpendicular to d. When $i_1 \neq i_2$, we may get cases in which only one of the yield lines is located over the support. The trapezoidal yield pattern thus formed can be dealt with by means of the formulae in section 2.23. In the case of simple supports, the reactions R_1 and R_2 lie on a line perpendicular to b and through the midpoint of the slab and not, as might be supposed, on a line parallel with a. They act within the support when the angle between b and d is acute, i.e. when m is determined by l. In the opposite case, the slab parts rotate about axes perpendicular to d, and the reactions act at the obtuse corners. In

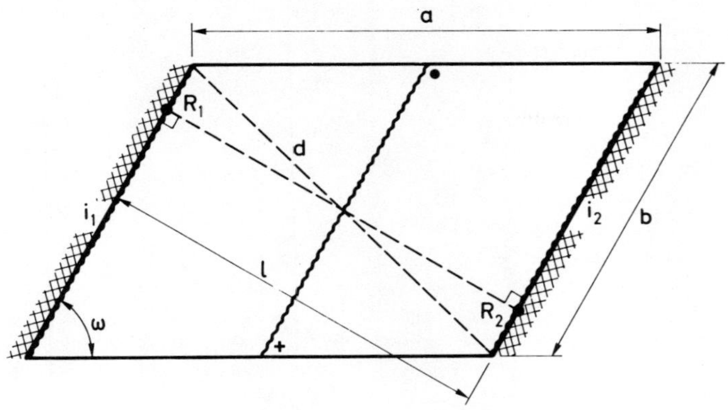

Fig. 2.121

this case we can obtain the more favourable yield pattern shown in Fig. 2.121 by anchoring the acute corners and placing top reinforcement at these. In accordance with the above rule, R acts at a distance $z = (1/2)\cot\omega - b/2$ outside the support. The anchorage force H is then determined by means of $Hb = Rz$, where $R = \frac{1}{2} pbl$. From this we get

$$H = \frac{1}{4} pl^2\left(\cot\omega - \frac{b}{l}\right) \qquad (2.122)$$

after which the formulae on pages 84-85 of YLT give

$$m' = \frac{1}{2} H \cot\frac{\omega}{2} = m\left(\cot\omega - \frac{b}{l}\right)\cot\frac{\omega}{2} = im \qquad (2.123)$$

In the sections perpendicular to d, the top reinforcement $m'' = i'm$ must be such that $d_r \leqslant l$, i.e.

$$1 + i' \geqslant \frac{d^2}{l^2} = \left(\frac{b^2}{l^2} + \frac{1}{\sin^2\omega} - 2\frac{b}{l}\cot\omega\right)$$

which gives

$$i' \geqslant \left(\cot\omega - \frac{b}{l}\right)^2$$

which is always less than i. If the top reinforcement is limited by lines through the obtuse corners, these lines must be drawn so that the distance between them is l.

For a point load P, we easily get (cf. example 0.21):

$$m = \frac{Pl_r^2}{4bl} = \frac{Pl}{b(\sqrt{1+i_1} + \sqrt{1+i_2})^2} \qquad (2.124)$$

when the yield pattern is as shown in Fig. 2.121. When the negative yield lines are not located along the supports, but form an angle with these (see Fig. 2.122), the same formula can be used as long as we insert b′ and l′ from the parallelogram AB′CD′ determined by these yield lines. It is then a case of determining the directions of b′ and l′ that will make l′/b′ maximum or will make b′/l′ minimum, which is the same thing. If we draw a circle of radius 1 about C, b′/l′ will be equal to the length of the tangent t to this circle between a and the short diagonal d. It will now be seen that the shortest of the tangents to the arc is the one that lies symmetrically, i.e. the one for which l′ bisects the angle between a and d, and then the yield line b′ bisects the obtuse angle between d and the extension of a. If we draw a line parallel with l′ through A (see Fig. 2.123), we get CS = CD′ = d, KS = $a + d - b\cos\omega$ and AK = $b\sin\omega$. Further,

$$\frac{l'}{b'} = \frac{1}{2\tan\theta} = \frac{KS}{2 \times AK} = \frac{a + d - b\cos\omega}{2b\sin\omega}$$

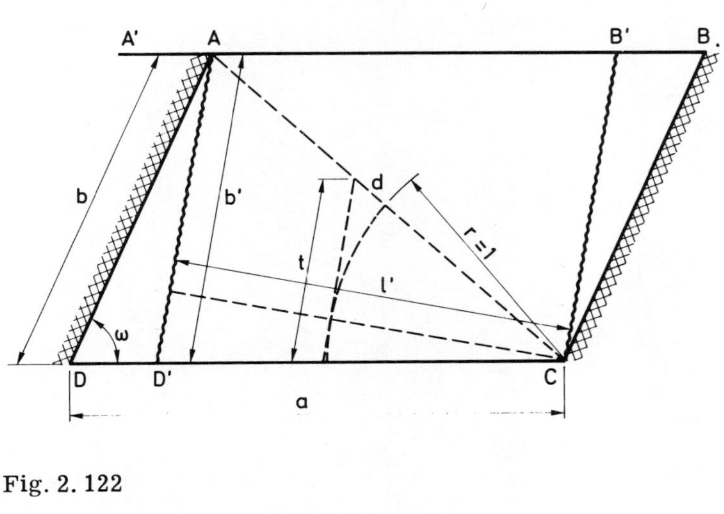

Fig. 2.122

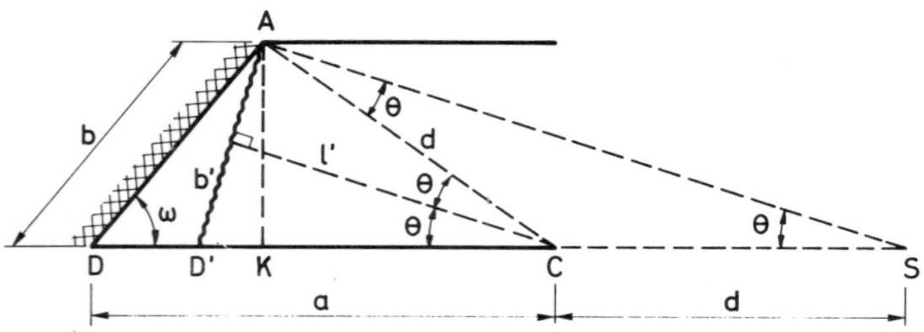

Fig. 2. 123

This finally gives

$$m = \frac{P}{2(\sqrt{1+i_1} + \sqrt{1+i_2})^2 \sin\omega}\left(\frac{a+d}{b} - \cos\omega\right) \quad (2.125)$$

This should only be used if D′ is located inside the support, i.e. $a > d$. It is here assumed that P acts near the mid-line parallel with a. When this is not the case, or when $i_1 \neq i_2$, we may get yield patterns with one yield line along the support. These cases can be dealt with using the formulae for trapezoidal slabs given in section 2. 23.

In the case of simple supports, the reactions R_1 and R_2 act on a line through P, which forms an angle $-\omega$ with b (see Fig. 2. 124). When $a < d$, the reactions are located inside the supports; when $a > d$, they are located outside, but then the slab rotates about the axes b′ (Fig. 2. 123) and the reactions act only at the obtuse corners; m is determined by means of (2. 125). By anchoring the acute corners, we can produce the yield pattern in Fig. 2. 121, and m is then determined by means of (2. 124). As the reaction acts a distance $z = l \cot\omega - \frac{1}{2}b$ outside the support, we get the anchorage force H and the negative moment in the same way as before:

$$H = \frac{Pl}{4b}\left(2\cot\omega - \frac{b}{l}\right)$$

$$m' = m\left(\cot\omega - \frac{b}{2l}\right)\cot\frac{\omega}{2} \quad (2.126)$$

The limit for the top reinforcement can be a line through the obtuse corner, such that $l'/b' = l/b$. It will immediately be seen from Fig. 2. 122 that t must have the same length as that corresponding to b, i.e. l' and l must be symmetrical about the bisector of the angle ACD, and consequently, b and b′ must be symmetrical about the bisector of the angle A′AC, i.e. b′ must form the angle ω with the short diagonal d.

If the slab is very wide, m cannot be less than the value corresponding to $P = 2\pi(m + m')$ for the circular yield pattern.

2.2 Trapezoidal slab

2.21 Supported on all four sides

We will find an approximate formula by means of the work equation and will use as yield pattern that determined by the bisectors and the mid-line, which gave very good results in the case of rectangular slabs. As all slab parts get the same rotation, this is reckoned to be equal to 1. The work equation then gives

$$m(1+i_1)a + m(1+i_2)b + m(1+i_3)c + m(1+i_4)d = \tfrac{1}{6}pr^2(a + b + 2s + c + d + 2s)$$

and as $2s = b + d - a - c$, we get

$$m(a + b + c + d) + m(i_1 a + i_2 b + i_3 c + i_4 d) = \tfrac{1}{6}pr^2(3b + 3d - a - c)$$

or

$$m + m\frac{i_1 a + i_2 b + i_3 c + i_4 d}{a + b + c + d} = m + m_i$$

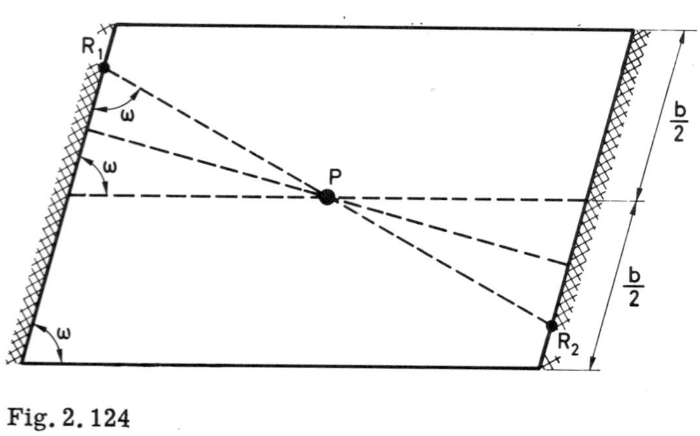

Fig. 2. 124

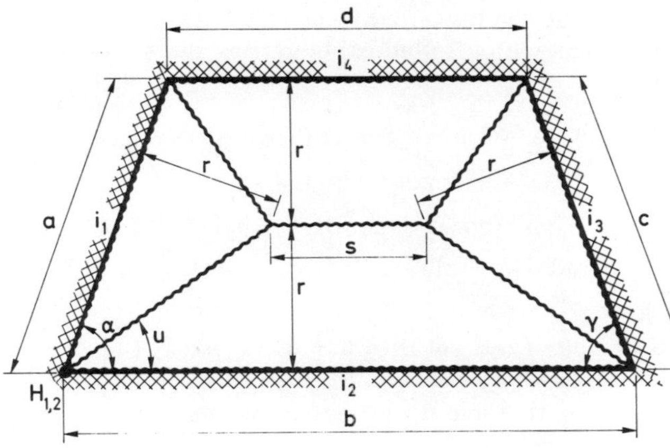

Fig. 2.211

$$= \tfrac{1}{6}pr^2 \frac{3\dfrac{b+d}{a+c} - 1}{\dfrac{b+d}{a+c} + 1} = m_o \qquad (2.211)$$

where the right side is the moment in a rectangular slab with the same height 2r as that of the given slab and the length

$$2r\,\frac{b+d}{a+c} < \frac{b+d}{2}$$

It is thus on the safe side to use the mean value. The influence of the corner levers is exactly the same as in the previous slabs, as both types have two acute and two obtuse angles. Corresponding to (2.112) we thus get

$$k = 1\cdot 75 - \tfrac{1}{3}(\sin\alpha + \sin\gamma) \qquad (2.212)$$

As the slab parts are triangles or trapezia, the reactions will still be uniformly distributed. As in the case of rectangular slabs, they can be reckoned to be equal to $\bar{q}\sqrt{1+i_1}$, $\bar{q}\sqrt{1+i_2}$, etc. For the corner force $H_{1,2}$, we have, for example,

$$H_{1,2} = m(1+i_1)\cot(\alpha - u) + m(1+i_2)\cot u$$

which is minimum for

$$\frac{1+i_1}{\sin^2(\alpha - u)} = \frac{1+i_2}{\sin^2 u}$$

or

$$\cot u = \cot\alpha + \sqrt{\frac{1+i_1}{1+i_2}}\,\frac{1}{\sin\alpha}$$

$$\cot(\alpha - u) = \cot\alpha + \sqrt{\frac{1+i_2}{1+i_1}}\,\frac{1}{\sin\alpha}$$

As the correct yield lines do not deviate much from the value thus determined, and as the variation at a minimum value is slight, this can be used to a good approximation. We thus get

$$H_{1,2} = \frac{2m\sqrt{(1+i_1)(1+i_2)}}{\sin\alpha} + (2 + i_1 + i_2)\,m\cot\alpha \qquad (2.213)$$

and correspondingly for the other corner forces. When these have been determined, the uniformly distributed reaction is found from the equation of projection:

$$\bar{q}(a\sqrt{1+i_1} + b\sqrt{1+i_2} + c\sqrt{1+i_3} + d\sqrt{1+i_4}) = P + \sum H \qquad (2.214)$$

As in the case of the previous slabs, more dangerous yield patterns may arise, the negative yield lines of which are not located along a and c, but at an angle with these (see Fig. 2.212). If a′, c′ are these yield lines, (2.211) must be used, with a′, b′, c′, d instead of a, b, c, d, and u and v are found by means of the maximum principle. If the height is h, we find that

$$\frac{d+b'}{a'+c'} = \frac{2\dfrac{d}{h} + \tan u + \tan v}{\dfrac{1}{\cos u} + \dfrac{1}{\cos v}} = \frac{1}{\sin u} = \frac{1}{\sin v} \qquad (2.215)$$

where the last terms are obtained by means of the condition for a maximum, (40) in YLT, page 64. We find that $\tan u = \tan v = h/d$, i.e. a′ and c′ are perpendicular to the diagonals in the rectangle hd, the diagonal l of which is exactly the length in the rectangular slab lh, which has the same moment as the given slab. The yield pattern only arises if a′ and c′ are located within a and c. If this is not the case with a, but is so with c, only the first equation in (2.215) is used for determination of u, as v is now

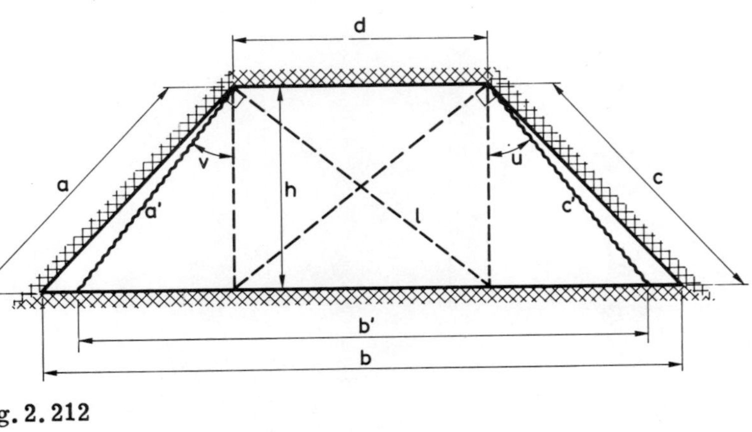

Fig. 2.212

given. As the length for determination of the moment is $l = h\,(d + b')/(a + c') = (h/\sin u)$, we find in this case that

$$l = h\,\frac{1 + A^2}{A \sec v + \sqrt{A^2 - \tan^2 v}}$$

where

$$A = 2\frac{d}{h} + \tan v \qquad (2.216)$$

In the special case of $v = 0$, we get

$$\tan\frac{u}{2} = \frac{1}{2}\,\frac{h}{d}, \quad l = d + \frac{h^2}{4d}$$

In the latter calculations, the fact that m_i is also dependent upon a', b', c' when the degrees of restraint are unequal is neglected. However, the investigation is mostly of theoretical interest as it does not apply at all when the mean width $\frac{1}{2}(b + d)$ is used in the calculation.

When the height is great, we get triangular slab parts at b and d and trapezoidal parts at a and c. Exact calculations confirm the obvious assumption that the slab should be calculated as a rectangle with the same height and the length $\frac{1}{2}(b + d)$. Such calculations also show that the procedure is slightly on the safe side (cf. also Ingerslev's rule discussed in section 2.3). The corner forces and reactions are calculated as above, as is also the influence of the corner levers. The treatment of triangular loads, line loads and point loads presents no difficulties either.

2.22 *Supported on three sides*

If we recall our investigations in 1.2, we will immediately see that the formulae derived there, (1.217) and (1.218), together with (1.223) and (1.224) for the rectangular slab with the sides d and h, can be used directly provided the corresponding yield pattern falls within the slab. This is always the case when the free edge d is shorter than b, but when d is longer than b, it is only the case when a and c are not too skew. When the yield pattern is not contained within the slab, new formulae have to be derived. We will here limit ourselves to simple approximate formulae derived by means of the work equation.

For uniformly distributed load p and line load $\bar{p}$ on the free edge, the work equation for the yield pattern formed by the bisectors (see Fig. 2.221) gives the following, with all slab parts getting the same rotation:

$$2mh\cot\alpha + 2mh\cot\beta + am_1' + bm_2' + cm_3'$$
$$= \tfrac{1}{6}pah^2 + \tfrac{1}{6}pch^2 + \tfrac{1}{2}pbh^2$$
$$- \tfrac{1}{3}ph^3(\cot\alpha + \cot\beta) + \tfrac{1}{2}\bar{p}ah + \tfrac{1}{2}\bar{p}ch$$
$$+ \bar{p}(d - a - c)h$$

where

$$h(\cot\alpha + \cot\beta) = a + b + c - d$$

Inserting this and rearranging the equation, we get

$$m + m_i = m_o$$
$$= \frac{1}{12}\,ph^2\,\frac{b + 2d - a - c}{a + b + c - d} + \frac{1}{4}\,\bar{p}h\,\frac{2d - a - c}{a + b + c - d} \qquad (2.221)$$

where

$$m_i = \frac{am_1' + bm_2' + cm_3'}{2(a + b + c - d)}$$

As shown in section 1.22, a triangular load that increases from zero at the free edge to p at b gives the same moment as the uniformly distributed load $\frac{1}{2}p$ and the line load $\bar{p} = -(1/12)ph$.

This immediately gives the following equation for the triangular load:

$$m_o = \frac{ph^2}{48}\,\frac{2b + 2d - a - c}{a + b + c - d} \qquad (2.222)$$

Dealing only with symmetrical cases, the work equation for the yield pattern in Fig. 2.222 gives

$$2m\left(a - \frac{d}{2}\cos\alpha\right) + mb + 2m_1'a + 2m_2'b$$
$$= 2 \times \frac{1}{6}p\left[\frac{d}{2\cos\alpha}\left(\frac{d}{2}\sin\alpha\right)^2 - \frac{b}{2\cos\alpha}\right.$$
$$\left.\times\left(\frac{b}{2}\cot\frac{\alpha}{2}\right)^2\right] + \frac{1}{6}pb\left(\frac{b}{2}\cot\frac{\alpha}{2}\right)^2 + 2\bar{p}\frac{d}{2}\frac{d}{4}\sin\alpha,$$

$$(d = b + 2a\cos\alpha)$$

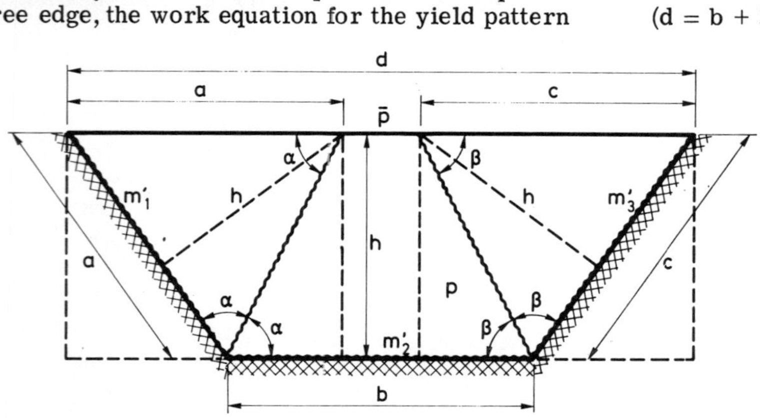

Fig. 2.221

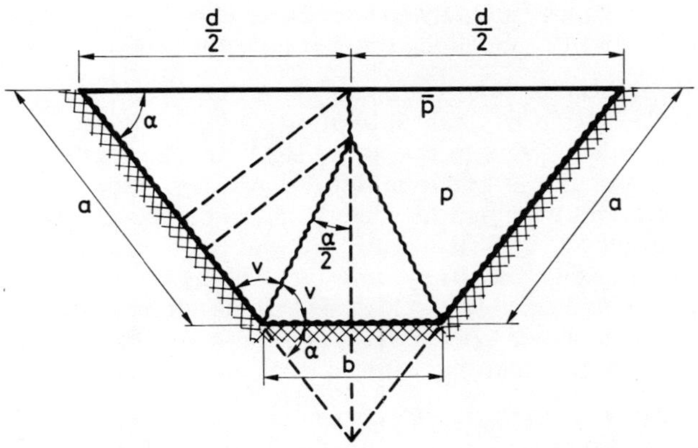

Fig. 2.222

After rearrangement we get

$$m + m_i = m_o$$
$$= \frac{p}{24}\,\frac{1+\cos\alpha}{1-\cos\alpha}\,\frac{2a(b^2+bd+d^2)-d^3}{2a+d}$$
$$+\frac{1}{4}\,\bar{p}\,\frac{d^2\sin\alpha}{(2a+d)\,(1-\cos\alpha)} \qquad (2.223)$$

with

$$m_i = \frac{2m'_1 a + m'_2 b}{(2a+d)\,(1-\cos\alpha)}$$

As a result of the long free edge, these slabs may be rather flexible.

Numerical example

A water canal with a depth of 2 m, a width of 2 m at the bottom, and a 1 : 1 slope of the sides, is limited by a vertical reinforced concrete slab. With $a = c = 2\sqrt{2}$, $b = 2$, $d = 6$ and $p = 2\ t/m^2$, we get, by means of (2.222),

$$m = \frac{2\times 4}{48}\,\frac{4(4-\sqrt{2})}{4(\sqrt{2}-1)} = 1{\cdot}04\ t$$

With the stresses 60 and 1000 kg/cm², we get a slab thickness of 12 cm and the deflexion

$$u = \frac{1}{8}\,\frac{1040\times 600^2}{2\times 10^5\times\frac{1}{12}\times 12^3} = 1{\cdot}6\ \text{cm}$$

$$> \frac{600}{500} = 1{\cdot}2\ \text{cm}$$

The thickness must be increased to 14 cm.

2.23 Supported on two sides

It seems natural to assume that this slab can be calculated as a rectangular slab with the same height h and length equal to the mean span s, which, with the notation in Fig. 2.231, is as follows:

$$s = (l + \tfrac{1}{2}h)\,(\tan\alpha + \tan\beta)$$

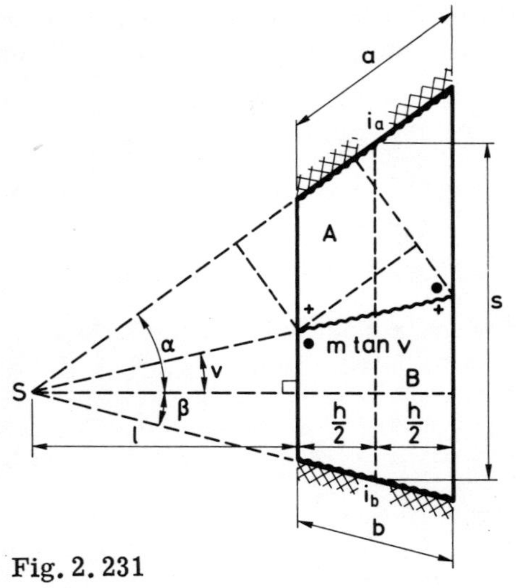

Fig. 2.231

As a correct solution to the problem presents no difficulties, the accuracy of this assumption can easily be investigated.

We write the moment equation about the support for slab part A and find

$$i_a m\,\frac{h}{\cos\alpha} + m\,\frac{h}{\cos v}\cos(\alpha - v)$$
$$+\, m\tan v\,\frac{h}{\cos v}\sin(\alpha - v)$$
$$= \frac{1}{6}p\,\frac{l+h}{\cos\alpha}\left[\frac{l+h}{\cos v}\sin(\alpha - v)\right]^2$$
$$-\frac{1}{6}p\,\frac{l}{\cos\alpha}\left[\frac{l}{\cos v}\sin(\alpha - v)\right]^2$$

or

$$m\left[\left(1+i_a\right)\frac{\cos^2 v}{\sin^2(\alpha - v)} - 1\right]$$
$$= \frac{1}{6}\,p\,(3l^2 + 3lh + h^2) \qquad (1)$$

We find a corresponding equation for slab part B, with i_b, β and $-v$ instead of i_a, α and v, respectively. The right side thus remains unaltered. The two left sides then give the following equation for determination of v:

$$\frac{\sin(\alpha - v)}{\sqrt{1+i_a}} = \frac{\sin(\beta + v)}{\sqrt{1+i_b}} \qquad (2)$$

which is easily solved, i.e.

$$\tan\left(v - \frac{\alpha-\beta}{2}\right) = \frac{\sqrt{1+i_b}-\sqrt{1+i_a}}{\sqrt{1+i_b}+\sqrt{1+i_a}}\tan\frac{\alpha+\beta}{2} \qquad (2.231)$$

It will be seen that $v - (\alpha - \beta)/2$ is the angle of the yield line with the bisector for $\alpha + \beta$. When $i_a = i_b$, the yield line therefore coincides with the bisector.

From (2) we now find, by dividing by $\cos\alpha\cos v$ and $\cos\beta\cos v$ in the numerator and denominator, that

$$\frac{\sin(\alpha - v)}{\sqrt{1+i_a}\cos v} = \frac{\tan\alpha - \tan v}{\sqrt{1+i_a}\sec\alpha} = \frac{\tan\beta + \tan v}{\sqrt{1+i_b}\sec\beta} = \frac{\tan\alpha + \tan\beta}{\sqrt{1+i_a}\sec\alpha + \sqrt{1+i_b}\sec\beta} \quad (3)$$

By this means we immediately get the following from (1) for the determination of m:

$$m\left[\left(\frac{\sqrt{1+i_a}\sec\alpha + \sqrt{1+i_b}\sec\beta}{\tan\alpha + \tan\beta}\right)^2 - 1\right] = \frac{1}{2}p\left(l^2 + lh + \frac{1}{3}h^2\right) \quad (2.232)$$

For simple support we get the special case of $i_a = i_b = 0$ and

$$m = \frac{1}{8}p\left(l^2 + lh + \frac{1}{3}h^2\right)(\tan\alpha + \tan\beta)^2 \times \frac{\cos\alpha\cos\beta}{\cos^2\frac{\alpha+\beta}{2}} \quad (2.233)$$

The approximate value is

$$m_t = \tfrac{1}{8}ps^2 = \tfrac{1}{8}p(l^2 + lh + \tfrac{1}{4}h^2)(\tan\alpha + \tan\beta)^2$$

$$\frac{m_t}{m} = \frac{l^2 + lh + \frac{1}{4}h^2}{l^2 + lh + \frac{1}{3}h^2}\,\frac{\cos^2\frac{\alpha+\beta}{2}}{\cos\alpha\cos\beta}$$

With a high l/h ratio, i.e. when the trapezium is fairly elongated, the first factor will be only slightly smaller than 1, and so the suitability of the approximation depends only upon the last term. This is exactly 1 when $\alpha = \beta$, but is otherwise greater, i.e. the approximation is on the safe side. Thus we get 20% for $\alpha = 45°$ and $\beta = 0$. Even greater degrees of error are found when β is negative, i.e. when the trapezium approaches the shape of a parallelogram. Thus, for $\alpha = 45°$ and $\beta = -15°$, we get a value that is 37% too high, and for $\alpha = 45°$ and $\beta = -30°$, a value that is 60% too high. In such cases it will be necessary to use the exact formulae.

In the special case of $i_a = i_b = i$ and $\alpha = \beta$, we get symmetry and

$$m = \tfrac{1}{2}p(l^2 + lh + \tfrac{1}{3}h^2)\,\frac{\tan^2\alpha}{1 + i\sec^2\alpha}$$

$$m_t = \tfrac{1}{2}p(l^2 + lh + \tfrac{1}{4}h^2)\,\frac{\tan^2\alpha}{1 + i'} \quad (2.234)$$

Neglecting the insignificant difference in the parentheses, we get the same value when the corresponding rectangular slab is given the restraints $i' = i\sec^2\alpha$, i.e. the oblique restraints have a significant effect.

In certain cases, the negative yield lines are not located along the supports but at an angle to these. In such cases, the same formula is used, but with α', β' and l' corresponding to the yield lines.

Between these quantities, the condition $l'(\tan\alpha' + \tan\beta') = l(\tan\alpha + \tan\beta)$ applies as the shorter of the parallel sides is the same for both yield patterns. α' and β' are then determined by means of the maximum principle. As these calculations are complicated, we will not go into them in detail but will just note that, in symmetrical slabs ($\alpha = \beta$), these cases only occur for restrained slabs with rather skew ends because less skew yield lines give a smaller restraining effect.

We will now turn to the treatment of a point load moving at a distance c from the shorter of the parallel sides. The load is distributed with P_A to slab part A and P_B to slab part B. In the moment equations formulated earlier it is only necessary to alter the load terms, which become

$$P_A(1 + c)\frac{\sin(\alpha - v)}{\cos v} \text{ and } P_B(1 + c)\frac{\sin(\beta + v)}{\cos v}$$

respectively, and thence

$$P_A = \frac{mh}{1+c}\left[\left(1 + i_a\right)\frac{\cos v}{\sin(\alpha - v)} - \frac{\sin(\alpha - v)}{\cos v}\right]\frac{1}{\cos\alpha}$$

and

$$P_B = \frac{mh}{1+c}\left[\left(1 + i_b\right)\frac{\cos v}{\sin(\beta + v)} - \frac{\sin(\beta + v)}{\cos v}\right]\frac{1}{\cos\beta}$$

By addition we then find $P = P_A + P_B$. The most dangerous location of P is obtained when $dP/dv = 0$, which gives (2) again, so we get the same yield line as before. Finally, by means of the expressions in (3), we get

$$P_A = \frac{mh}{1+c}\,\frac{\sqrt{1+i_a}}{\cos\alpha}K, \quad P_B = \frac{mh}{1+c}\,\frac{\sqrt{1+i_b}}{\cos\beta}K$$

$$P = \frac{mh}{1+c}\left(\frac{\sqrt{1+i_a}}{\cos\alpha} + \frac{\sqrt{1+i_b}}{\cos\beta}\right)K$$

where

$$K = \frac{\sqrt{1+i_a}\sec\alpha + \sqrt{1+i_b}\sec\beta}{\tan\alpha + \tan\beta} - \frac{\tan\alpha + \tan\beta}{\sqrt{1+i_a}\sec\alpha + \sqrt{1+i_b}\sec\beta} \quad (2.235)$$

i.e.

$$P = \frac{mh}{(1+c)(\tan\alpha + \tan\beta)} \times \left[(\sqrt{1+i_a}\sec\alpha + \sqrt{1+i_b}\sec\beta)^2 - (\tan\alpha + \tan\beta)^2\right]$$

For simple support, we get the special case of $i_a = i_b = 0$, and thus

$$m = \frac{P(1+c)}{4h}\,2\tan\frac{\alpha+\beta}{2} \quad (2.236)$$

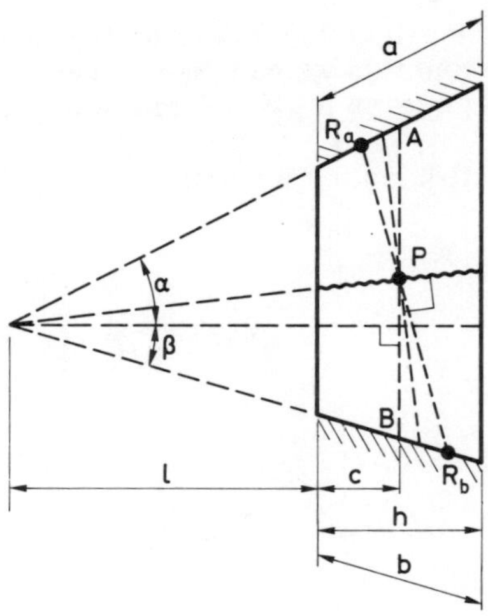

Fig. 2.232

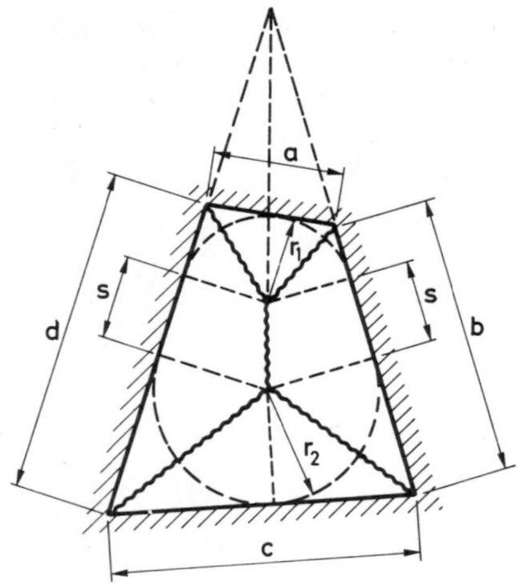

Fig. 2.31

which is the same as in a rectangular slab with the span $2(l + c) \tan(\alpha/2 + \beta/2)$. The span is thus found by drawing a line perpendicular to the bisector of $\alpha + \beta$. The reactions on the sides a and b are $R_a = P_A$ and $R_b = P_B$, which are proportional to the sides on which they act. Their distance from P must be proportional to $\cos\alpha$ and $\cos\beta$. We thereby arrive at the determination of the location of R_a and R_b shown in Fig. 2. 232. We draw a line through P, symmetrical to AB with regard to the bisector. The maximum moment occurs when $c = h$, and R_b then acts outside the support. With a symmetrical yield pattern, both R_a and R_b act at the corners, i.e. slab part B (in Fig. 2. 231) rotates about a line symmetrical to side a through the outermost corner. We then get

$$m = \frac{P}{4}\frac{l' + h}{h} 2 \tan\alpha \qquad (2.236')$$

Finally, in the symmetrical case, $i_a = i_b = i$, $\alpha = \beta$, we get

$$m = \frac{P(l + c)}{4h} \frac{2 \tan\alpha}{1 + i \sec^2\alpha} \qquad (2.237)$$

which again shows that the oblique restraints act with $i \sec^2\alpha$.

2.3 Convex quadrilateral slab supported on four sides

This type of slab has already been dealt with by Ingerslev in a paper in "Ingeniøren", 1921, page 510, which was of fundamental importance for the yield-line theory. If we use the approximation, $m + m_i = m_o$, where m_o is the moment for simple support and m_i the mean restraint moment, we will only have to determine m_o. For the special case of a circumscribed quadrilateral, with the radius r of the inscribed circle, we get (YLT, page 60), $m = \frac{1}{6}pr^2$ for uniformly distributed load. This has to be corrected for the influence of the corner levers, which is done in the same way as for the triangular slabs in the next chapter.

In the general case, we use the yield pattern formed by the bisectors (Fig. 2. 31). All slab parts thereby get the same rotation, which is put at 1, and the work equation then gives

$$m(a + b + c + d) = 2 \times \frac{1}{6} p a r_1^2 + 2 \times \frac{1}{6} p c r_2^2 + 2s \times \frac{1}{6} p(r_1^2 + r_1 r_2 + r_2^2) \qquad (2.31)$$

where r_1 and r_2 are the radii of the two tangential circles for b, d and a or c. We have $2s + a + c = b + d$.

In accordance with Ingerslev, we determine the point of intersection 0 between the bisectors of the angles formed by opposite sides (Fig. 2. 32). The distance of this point from the sides is r and R, respectively. The moment is then the same as in a rectangular slab with the sides 2r and 2R. This approximation is acceptable as long as the projections of 0 on the

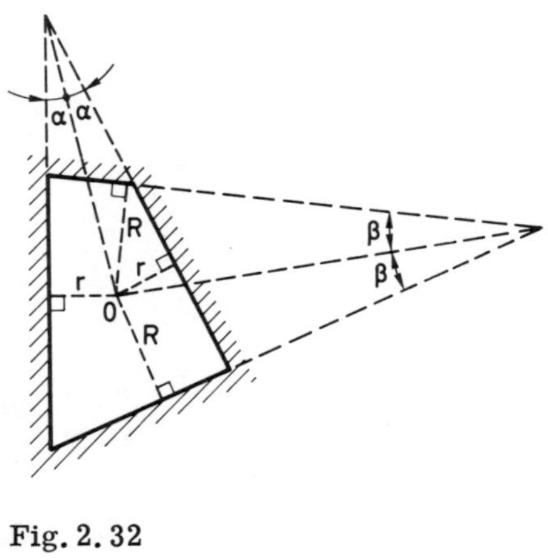

Fig. 2. 32

sides are located on the periphery of the quadrilateral. For a parallelogram and for a symmetrical trapezium, this method of calculation coincides with that specified earlier. In these cases we get the degenerate yield patterns (Figs. 2.112 and 2.212) precisely when the projection of 0 is located outside the periphery. From this we can take it that such degenerate yield patterns must be used in this case.

Formula (1.14) still applies for point loads.

3 Triangular slabs

These slabs are more of theoretical than of practical interest, but they are encountered in reinforced concrete structures in the form of wing walls at bridge piers and in steel structures in the form of base-plates.

A special characteristic of these slabs is the significance of the corner levers.

3.1 Supported on three sides

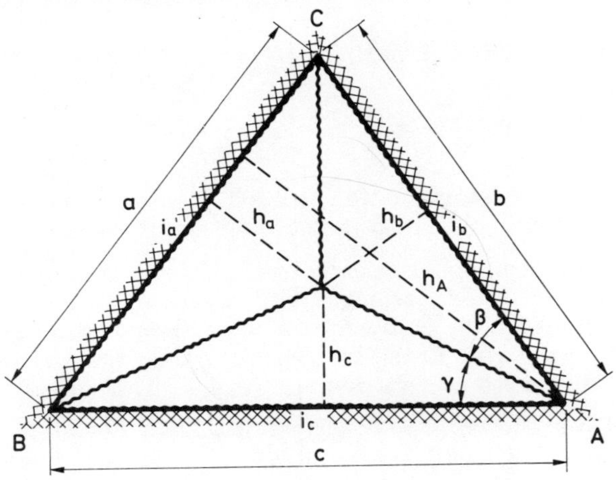

Fig. 3.11

When the degrees of restraint on the sides a, b and c are i_a, i_b and i_c, respectively, the moment equations about the sides are as follows in the case of a uniformly distributed load:

$$m(1 + i_a) = \frac{1}{6} ph_a^2, \quad m(1 + i_b) = \frac{1}{6} ph_b^2,$$

$$m(1 + i_c) = \frac{1}{6} ph_c^2$$

Further, $ah_a + bh_b + ch_c = 2T$, where T is the area of the triangle. Introducing here the h's expressed by means of the first three equations, we get, with $P = pT$,

$$m = \frac{2}{3} P \frac{T}{(a\sqrt{1 + i_a} + b\sqrt{1 + i_b} + c\sqrt{1 + i_c})^2}$$

$$= \frac{P}{3} \frac{\sin A \sin B \sin C}{(\sqrt{1 + i_a} \sin A + \sqrt{1 + i_b} \sin B + \sqrt{1 + i_c} \sin C)^2} \qquad (3.11)$$

In particular, when $i_a = i_b = i_c$, we get

$$m + m' = \frac{2}{3} P \frac{T}{(a + b + c)^2} = \frac{1}{6} pr^2$$

where r is the radius of the inscribed circle.

In accordance with YLT (108), page 138, the reaction on side a is

$$\bar{q}_a = \frac{2}{3} ph_a = \frac{4}{3} P \frac{\sqrt{1 + i_a}}{a\sqrt{1 + i_a} + b\sqrt{1 + i_b} + c\sqrt{1 + i_c}} \qquad (3.12)$$

and correspondingly for the other sides. When the i's are equal, we get the same value of $\bar{q}$ on all three sides.

At the corner at A, we get the corner force

$$H_A = m(1 + i_c) \cot \gamma + m(1 + i_b) \cot \beta$$

where $\beta + \gamma = A$ and $\sin \beta / \sin \gamma = \sqrt{1 + i_b}/\sqrt{1 + i_c} = h_b/h_c$. From $\sin \beta = \sin (A - \gamma)$, we find that

$$\cot \gamma = \frac{\sqrt{1 + i_b} + \sqrt{1 + i_c} \cos A}{\sqrt{1 + i_c} \sin A}$$

and

$$\cot \beta = \frac{\sqrt{1 + i_c} + \sqrt{1 + i_b} \cos A}{\sqrt{1 + i_b} \sin A}$$

When these are inserted, we get

$$H_A = \frac{m}{\sin A} [2\sqrt{1 + i_b}\sqrt{1 + i_c} + (2 + i_b + i_c) \cos A] \qquad (3.13)$$

It will be seen that H_A is exactly the minimum value in (2.213).

As $2T = bc \sin A$ and $a^2 = b^2 + c^2 - 2bc \cos A$, sin A and cos A can hereby be expressed by means of the sides, and we find

$$H_A = \frac{m}{4T} [4b\sqrt{1 + i_b}\, c\sqrt{1 + i_c} + (2 + i_b + i_c)(b^2 + c^2 - a^2)] \qquad (3.13')$$

and corresponding expressions for H_B and H_C. We find that $H_A + H_B + H_C = \frac{1}{3}P$. Further, $\bar{q}_a a + \bar{q}_b b + \bar{q}_c c = \frac{4}{3}P$, so the check agrees as the difference is exactly P.

When the i's are equal, we get, in particular,

$$H_A = \frac{m + m'}{2T} (a + b + c)(b + c - a) \qquad (3.13'')$$

Because of the acute angles it is not possible to reinforce the top sufficiently heavily to avoid corner levers, and as these are of such importance here, they must be investigated in detail. However, we will assume that $i_a = i_b = i_c = i$ and will not reinforce the corners more strongly than is required for i.

As all slab parts experience the same rotation when $h_a = h_b = h_c = r$, we obtain

$$(m + m')(a + b + c) - \triangle A - \triangle B - \triangle C = P\frac{r}{3}$$

$$= \frac{1}{6}pr^2(a + b + c)$$

as the corrected work equation, $\triangle A$, $\triangle B$ and $\triangle C$ being the corrections for the corners A, B and C. (We here assume that $\theta = 1$.) Denoting the height of the corner lever h, we get

$$m + m' = \frac{1}{6}ph^2$$

Further, as $\triangle A$ has the form $2h\,(m + m')\,V_A$ [cf. YLT, page 96, (69)], we have the following equation for the determination of h and thus of $m + m'$:

$$h^2 - 2h^3\frac{V_A + V_B + V_C}{a + b + c} = r^2$$

As the ratio between the moment that takes account of the corner levers and the moment neglecting these is $k = h^2/r^2$, we get the following for determination of the correction factor k:

$$k = 1 + \delta k\sqrt{k},\ \delta = \frac{2r}{a + b + c}(V_A + V_B + V_C) \tag{3.14}$$

which is easily solved by trial and error or by iteration since δ is small in comparison with 1. Moreover, comparative calculations show that we have, to a good approximation,

$$k = 2 - \sqrt{1 - 2{\cdot}8\delta} \tag{3.14'}$$

According to (69),

$$V_A = \sqrt{\frac{K}{K - 1}}\,\frac{2(K - 1)^2}{(\sqrt{1 + 3K})^3 + 9K - 1}$$

where K is given by (66) and $\theta_A = \theta_B = 1$. Further, m' in (66) must be assumed to be zero in accordance with the observation concerning the corner levers of restrained slabs in YLT, page 98.

Comparative calculations now show that this complicated expression can, to a good approximation, be replaced by the simpler expression,

$$V_A = \frac{1}{5}\cot\frac{A}{2}\left(1 - \sin\frac{A}{2}\right)\left(2 - \sin\frac{A}{2}\right) \tag{3.15}$$

δ and thus k can thereby be determined. A number of examples are given in the table on page 67.

It will be seen that the corner levers increase the moment by from 20 to 40%.

From the values for k, we find that the following is an acceptable approximation:

$$k = 1{\cdot}85 - \frac{1}{4}(\sin A + \sin B + \sin C) \tag{3.16}$$

which can, however, only be utilized in the case of triangles with angles that are equal to or greater than 30°.

In the case of a triangular load, with $p = 0$ along side a, and elsewhere $p = \kappa x$, where x is the distance from side a, we get for the slab part at a:

$$m(1 + i_a)a = \kappa I_a = \frac{\kappa}{12}ah_a{}^3$$

for the slab part at b (see Fig. 3. 11),

$$m(1 + i_b)b = \kappa Z_{ab} = \kappa\frac{1}{6}bh_b\left[\frac{1}{2}h_b \times \frac{1}{2}h_a + \frac{1}{2}h_b \times \frac{1}{2}(h_a + h_A)\right] = \frac{\kappa bh_b{}^2}{12}\left(h_a + \frac{1}{2}h_A\right)$$

and for the slab part at c,

$$m(1 + i_c)c = \frac{\kappa c}{12}h_c{}^2\left(h_a + \frac{1}{2}h_A\right)$$

For the determination of I and Z, we can use the following little-known but very useful theorem*:

Statical moments, centrifugal moments and moments of inertia for a triangle with regard to arbitrary orthogonal or oblique axes can be found by imagining that the area of the triangle is concentrated at the mid-points of its sides, with one-third at each.

From the last two equations it will immediately be seen that $h_b/h_c = \sqrt{1 + i_b}/\sqrt{1 + i_c}$. We therefore put

$$h_b = h\sqrt{\frac{1 + i_b}{1 + i_a}} \text{ and } h_c = h\sqrt{\frac{1 + i_c}{1 + i_a}}$$

and get

$$\frac{12}{\kappa}m(1 + i_a) = h^2\left(h_a + \frac{1}{2}h_A\right) = h_a{}^3 \tag{3.17}$$

Further, $ah_a + bh_b + ch_c = 2T = ah_A$, or

$$h_A - h_a = \frac{b\sqrt{1 + i_b} + c\sqrt{1 + i_c}}{a\sqrt{1 + i_a}}h = \alpha h$$

Inserting this in (3. 17) and with $P = \frac{1}{6}\kappa ah_A{}^2$ and $a = (\cot B + \cot C)\,h_a$, we find

$$\left(\frac{h_A}{h_a}\right)^3 = 3\frac{h_A}{h_a} + 2(\alpha^2 - 1)$$

$$\frac{P}{m(1 + i_a)} = 2(\cot B + \cot C)\left(\frac{h_A}{h_a}\right)^3 \tag{3.18}$$

$$\alpha = \frac{\sqrt{1 + i_b}\sin B + \sqrt{1 + i_c}\sin C}{\sqrt{1 + i_a}\sin A}$$

---

* cf., for example, Jacob Nielsen: Dynamik. Copenhagen, Jul. Gjellerup, 1945, page 189, example 16.

Table

Angles			P/(m + m′) for uniformly distributed load			P/(m + m′) for triangular loading, p = 0 at		
A	B	C	uncorrected	k	corrected	a	b	c
60°	60°	60°	31·2	1·20	26·0	30·0	30·0	30·0
90°	45°	45°	35·0	1·25	28·0	32·0	34·8	34·8
90°	60°	30°	38·8	1·30	30·0	35·4	36·7	41·5
120°	30°	30°	48·2	1·40	34·5	42·5	49·0	49·0

On the right in the above Table, the values of P/(m + m′) are given for the same slabs with triangular loading, where p = 0 at a, b and c, respectively.

A comparison with the values for uniformly distributed load shows that when the triangular loading is zero at the longest side, the moment is slightly higher (9 to 13%) than in the case of uniformly distributed load. When the triangular loading is zero at the shortest side, we get a slightly smaller moment. When correcting for the corner levers, we can use the same value of k as for uniformly distributed load.

For a point load, we get the same conditions as in 1.14, so formula (1.14) can still be used, although only up to n = 2. For n = 3, we now have $P = 2\pi(m + m')$.

When calculating the deflexions, we must select a suitable strip that includes the maximum deflexion and maximum moment.

It is obvious that this strip must pass through the intersection of the yield lines and must be approximately perpendicular to one of these. We thereby usually get three possibilities giving slightly different spans. This indeterminateness, which is really insignificant, can be avoided by discarding the straight strips and using, instead, torsionless strips along broken lines. For the equilateral triangle, the radii of the inscribed circle shown in Fig. 3.12 are axes of symmetry and therefore torsionless.

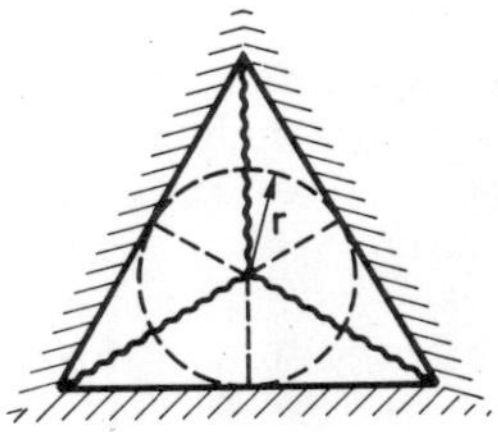

Fig. 3.12

The deflexion at the centre will then be the same as for a straight strip of length 2r. Assuming, as usual, that the moment is constant and equal to m, we get

$$EIu = \frac{1}{2} mr^2 \qquad (3.19)$$

With m = P/31·2, we thus get the exact deflexion for the equilateral triangle.

For an arbitrary triangle, the maximum deflexion occurs at N, where the tangent plane of the surface or deflexion is horizontal. The three strips shown therefore have the same deflexions as beams built in at N.

When $i_a = i_b = i_c$, we can, at any rate approximately, assume the same variation of the moment in the three strips, and as these have to give the same deflexion, they must be of equal length. Consequently, N will be the centre of the inscribed circle. With constant moment equal to m, we again get (3.19). When there are different degrees of restraint at the sides, we may still use (3.19) because we are only seeking an estimate of the deflexion.

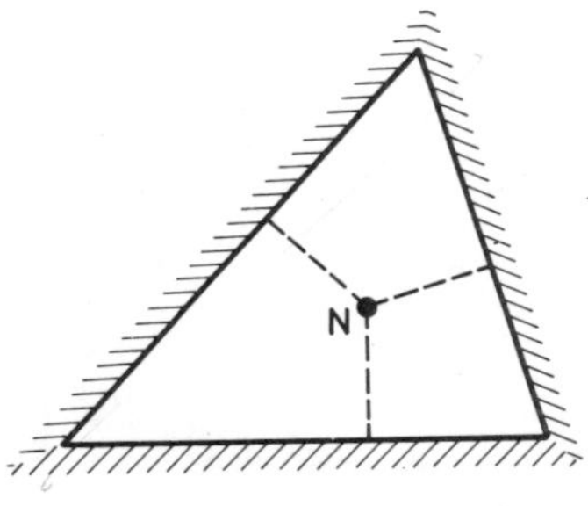

Fig. 3.13

Numerical example

A right-angled slab with sides of 6, 8 and 10 m is simply supported and subjected to a load of p = 600 kg/m². We get r = 2 and $m = \frac{1}{6} \times 600 \times 2^2 = 400$ kg. This must be corrected for the corner levers, and (3.16) gives k = 1·85 — 0·25(1 + 0·6 + 0·8) = 1·25, i.e. m = 500 kg. From this we get a slab thickness of 8 cm, which results in the following deflexion:

$$u = \frac{1}{2} \frac{500 \times 200^2}{2 \times 10^5 \times \frac{1}{12} \times 8^3} = 1{\cdot}17 \text{ cm}$$

$$> \frac{400}{500} = 0{\cdot}8 \text{ cm}$$

as the span is assumed to be 2r. To improve the

stiffness of the slab, the thickness must be increased to 10 cm.

3.2 Triangular slab supported on two sides

If it is assumed that $\theta_A = 1/h_a$ and $\theta_B = 1/h_b$ in the yield pattern shown in Fig. 3.21, the work equation gives

$$(ms + i_a ma)\frac{1}{h_a} + (mt + i_b mb)\frac{1}{h_b}$$

$$= \frac{1}{6}\, pah_a + \frac{1}{6}\, pbh_b = \frac{P}{3}$$

for uniformly distributed load. With triangular loading proportional to the distance from the free edge (and zero here), the right side becomes P/4, since the resultant of the triangular load lies at a distance of $\frac{1}{4}\, h_b(h_a)$ from b(a).

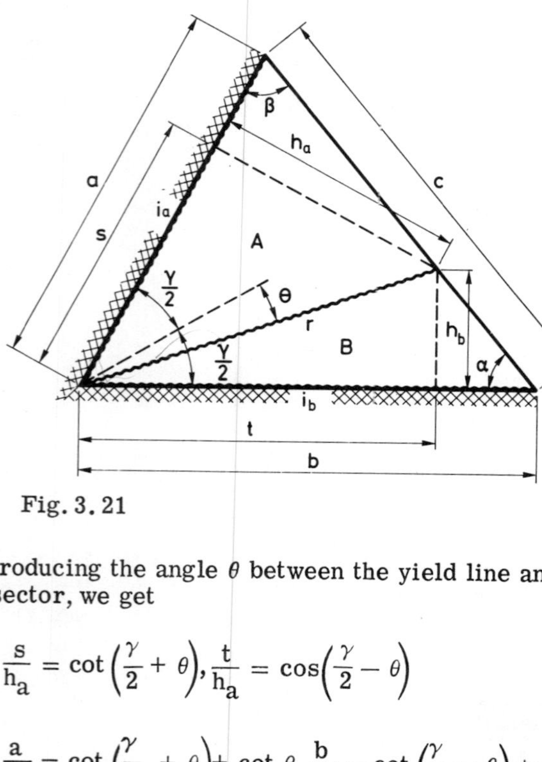

Fig. 3.21

Introducing the angle θ between the yield line and the bisector, we get

$$\frac{s}{h_a} = \cot\left(\frac{\gamma}{2} + \theta\right), \frac{t}{h_a} = \cos\left(\frac{\gamma}{2} - \theta\right)$$

$$\frac{a}{h_a} = \cot\left(\frac{\gamma}{2} + \theta\right) + \cot\beta, \frac{b}{h_b} = \cot\left(\frac{\gamma}{2} - \theta\right) + \cot\alpha$$

This immediately gives

$$(1 + i_a)\cot\left(\frac{\gamma}{2} + \theta\right) + (1 + i_b)\cot\left(\frac{\gamma}{2} - \theta\right)$$

$$+ i_a \cot\beta + i_b \cot\alpha = \frac{P}{3m} \qquad (3.21)$$

The maximum principle then gives

$$\frac{1 + i_a}{\sin^2\left(\frac{\gamma}{2} + \theta\right)} = \frac{1 + i_b}{\sin^2\left(\frac{\gamma}{2} - \theta\right)}$$

$$\frac{\sin\left(\frac{\gamma}{2} + \theta\right)}{\sin\left(\frac{\gamma}{2} - \theta\right)} = \sqrt{\frac{1 + i_a}{1 + i_b}} = \frac{h_a}{h_b} = \frac{\tan\frac{\gamma}{2} + \tan\theta}{\tan\frac{\gamma}{2} - \tan\theta}$$

$$\tan\theta = \frac{\sqrt{1 + i_a} - \sqrt{1 + i_b}}{\sqrt{1 + i_a} + \sqrt{1 + i_b}}\tan\frac{\gamma}{2} \qquad (3.22)$$

which, inserted in (3.21), gives

$$m = \frac{2P}{3}\,\frac{\tan\frac{\gamma}{2}}{(\sqrt{1 + i_a} + \sqrt{1 + i_b})^2 + 2\tan\frac{\gamma}{2}(i_a\cot\beta + i_b\cot\alpha) - \tan^2\frac{\gamma}{2}(\sqrt{1 + i_a} - \sqrt{1 + i_b})^2} \qquad (3.23)$$

When $i_a = i_b = i$, we get $\theta = 0$ and

$$m = \frac{P}{3\left[2(1 + i)\cot\frac{\gamma}{2} + i(\cot\alpha + \cot\beta)\right]} \qquad (3.23')$$

It should be noted that

$$\cot\alpha + \cot\beta = \left(\frac{a}{b} + \frac{b}{a}\right)\frac{1}{\sin\gamma} - 2\cot\gamma$$

In the special case of right-angled triangles, $\gamma = \pi/2$, we get

$$m = \frac{P}{3[2\sqrt{(1 + i_a)(1 + i_b)} + i_a\tan\alpha + i_b\cot\alpha]} \qquad (3.23'')$$

It should be noted that θ in (3.22) does not depend upon the ratio between the sides.

In the case of triangular loading with $p = 0$ at the free edge, P/3 is replaced by P/4, and in the case of line load on the free edge, it is replaced by P/2, since in this case the resultant load on A acts a distance $\frac{1}{2}h_a$ from A, and correspondingly for B. The three loading cases thus give the same yield pattern and can consequently be superposed without approximation. Thus, for a slab subjected to a uniformly distributed load P_e, a triangular load P_k and a line load $\overline{P}$, we can, for example, in (3.23), replace $\frac{2}{3}P$ by $\frac{2}{3}P_e + \frac{1}{2}P_k + \overline{P}$.

For the sake of simplicity we will only investigate the influence of the corner levers for a = b and $i_a = i_b = i$. Because of the symmetry, we then get the yield pattern shown in Fig. 3.22. The negative yield line is assumed to have the same ultimate moments as the restraints, i.e. $m' = im$.

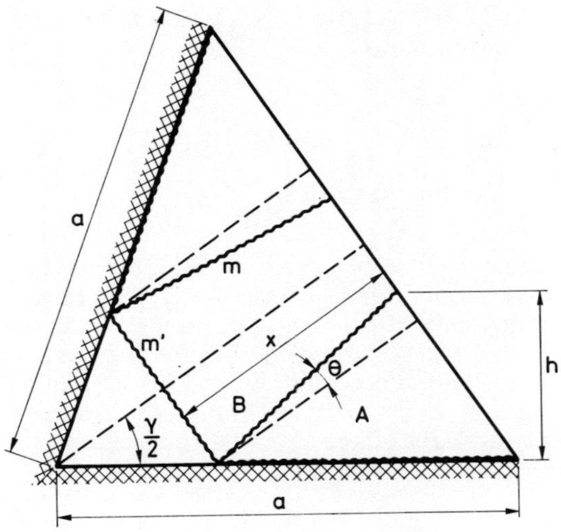

Fig. 3. 22

For a deflexion $\delta = 1$ at the middle of the free edge, we get

$$\theta_A = \frac{1}{h}, \theta_B = \frac{1}{x}$$

where

$$h = \frac{x}{\cos\theta} \sin\left(\theta + \frac{\gamma}{2}\right)$$

Then, with the contribution from A and half of B, written in the order mentioned, the work equation becomes

$$0 = \left[-m \frac{x}{\cos\theta} \cos\left(\theta + \frac{\gamma}{2}\right) - im \frac{x}{\cos\frac{\gamma}{2}} + \frac{1}{6} p \frac{x}{\cos\frac{\gamma}{2}} \frac{x^2}{\cos^2\theta} \sin^2\left(\theta + \frac{\gamma}{2}\right)\right] \frac{\cos\theta}{x \sin\left(\theta + \frac{\gamma}{2}\right)}$$

$$+ \left[-mx \tan\theta - im\left(a \sin\frac{\gamma}{2} - x \tan\frac{\gamma}{2}\right) + \frac{1}{2} px^2\left(a \sin\frac{\gamma}{2} - x \tan\frac{\gamma}{2}\right) - \frac{1}{3} px^3 \tan\theta\right] \frac{1}{x}$$

or, when rearranged,

$$m\left(\frac{\cos\frac{\gamma}{2}}{\cos\theta \sin\left(\theta + \frac{\gamma}{2}\right)} + i\left[\cot\left(\theta + \frac{\gamma}{2}\right) + \frac{a}{x} \sin\frac{\gamma}{2}\right]\right) = \frac{1}{2} pax \sin\frac{\gamma}{2} - \frac{1}{6} px^2\left(\tan\theta + 2 \tan\frac{\gamma}{2}\right) \tag{3.24}$$

We will first deal with the simply supported slab, i.e. $i = 0$. The maximum principle then immediately gives $\partial m/\partial x = 0$, thus:

$$x = \frac{3}{2} \frac{a \sin\frac{\gamma}{2}}{\tan\theta + 2 \tan\frac{\gamma}{2}}$$

$$m = \frac{3P}{8} \tan\frac{\gamma}{2} \frac{\tan\theta + \tan\frac{\gamma}{2}}{(1 + \tan^2\theta)\left(\tan\theta + \tan\frac{\gamma}{2}\right)} \tag{3.25}$$

As $\partial m/\partial \tan\theta = 0$, we then get from the latter equation

$$\tan^3\theta + \frac{5}{2} \tan\frac{\gamma}{2} \tan^2\theta + 2 \tan^2\frac{\gamma}{2} \tan\theta = \frac{1}{2} \tan\frac{\gamma}{2}$$

from which $\tan\theta$ can be determined. The ratio between the moment with corner levers and that without is

$$k = \frac{9}{4} \frac{\tan\theta + \tan\frac{\gamma}{2}}{(1 + \tan^2\theta)\left(\tan\theta + 2 \tan\frac{\gamma}{2}\right)} \tag{3.26}$$

The usual corner lever with its vertex on a yield line may also occur. Neglecting the corner levers, we have $m_o = \frac{1}{6} p h_o^2 = \frac{1}{6} P \tan(\gamma/2) = \frac{1}{6} pa^2 \sin^2(\gamma/2)$, so $h_o = a \sin(\gamma/2)$. With $\theta = 1$, the work equation, with the corner lever correction ΔA, gives

$$2ma \cos^2\frac{\gamma}{2} - \Delta A = 2 \times \frac{1}{6} paa^2 \cos^2\frac{\gamma}{2} \sin^2\frac{\gamma}{2}$$

where $\Delta A = 2mhV$ with the notation from section 3.1. Inserting $h = (h/h_o) a \sin(\gamma/2)$, $m = \frac{1}{6} ph^2$, we get

$$\left(\frac{h}{h_o}\right)^2 - \left(\frac{h}{h_o}\right)^3 \frac{\sin\frac{\gamma}{2}}{\cos^2\frac{\gamma}{2}} V = 1$$

or, by means of (3.15),

$$\delta = \frac{\sin\frac{\gamma}{2}}{\cos^2\frac{\gamma}{2}} V = \frac{1 - \sin\frac{\gamma}{2}}{5 \cos\frac{\gamma}{2}} \left(2 - \sin\frac{\gamma}{2}\right)$$

Corresponding to k we then get $k' = (h/h_o)^2$, which is found from the above equation, which corresponds exactly to (3.14). The calculations now show that, for $\gamma \geqslant 60°$, we get the corner levers first dealt with, and we can, with excellent accuracy, put

$$k = \frac{11 - 2 \sin\frac{\gamma}{2}}{8} \text{ for } \gamma \geqslant 60° \tag{3.271}$$

If we assume the same value when $a \neq b$, we get the following for simple support:

$$m = \frac{P}{48}\left(11 - 2 \sin\frac{\gamma}{2}\right) \tan\frac{\gamma}{2} \text{ for } \gamma \geqslant 60° \tag{3.272}$$

which applies for uniformly distributed load. This equation can also be used as an approximation for triangular loading and combinations of these and line loading when the factors from page 68 are inserted.

Now it is found that k′ is closely approximated by the following expression:

$$k' = 0{\cdot}9 \frac{\sin\frac{\gamma}{2} + \frac{1}{3}}{\sin\frac{\gamma}{2} + 0{\cdot}1} \text{ for } \gamma \leqslant 60° \qquad (3.273)$$

and we thus get

$$m = 0{\cdot}15\, P \frac{\sin\frac{\gamma}{2} + \frac{1}{3}}{\sin\frac{\gamma}{2} + 0{\cdot}1} \tan\frac{\gamma}{2} \text{ for } \gamma \leqslant 60° \qquad (3.274)$$

We will now turn to the restrained slab with i = 1.

In the calculation of the usual corner levers, we have $m + m' = 2m = \frac{1}{6} ph^2$. Without corner levers,

$$m_o = \frac{1}{6} pa^2 \frac{\sin^2\frac{\gamma}{2}}{2 + \tan^2\frac{\gamma}{2}} = \frac{1}{12} ph_o^2$$

i.e.

$$h_o = \frac{a \sin\frac{\gamma}{2}}{\sqrt{1 + \frac{1}{2}\tan^2\frac{\gamma}{2}}}$$

With $\theta = 1$, the work equation becomes

$$2ma \cos^2\frac{\gamma}{2} + ma - \Delta A$$

$$= 2 \times \frac{1}{6} paa^2 \cos^2\frac{\gamma}{2} \sin^2\frac{\gamma}{2}$$

where
$\Delta A = 2(m + m')hV = 4mhV$.
We then find the usual equation,
$(h/h_o)^2 - (h/h_o)^3\, \delta = 1$,
but now with

$$\delta = \frac{\sin\frac{\gamma}{2}}{\cos^2\frac{\gamma}{2}} V\Big/\left(1 + \frac{1}{2}\tan^2\frac{\gamma}{2}\right)^{3/2}$$

i.e. a considerably lower value than for i = 0.

Calculations for i = 1 similar to those above for i = 0 show that the transition from the usual corner lever to the special corner lever in Fig. 3.22 takes place at $\gamma = 90°$. The calculations further show that the maximum correction for $\gamma > 90°$ is 7%, but increases rapidly when $\gamma < 90°$. We can therefore assume that k is constant and equal to 1 for $\gamma \geqslant 90°$.

For $\gamma < 90°$, the calculations show that we have, to a good approximation,

$$k' = 0{\cdot}8 \frac{\sin\frac{\gamma}{2} + \frac{3}{8}}{\sin\frac{\gamma}{2} + 0{\cdot}1} \qquad (3.28)$$

For other degrees of restraint, we interpolate between the values for i = 0 and i = 1, just as the slab restrained only at one side is assumed to have the mean value for i = 0 and i = 1 on both sides. This applies to the correction factors k or k′, since the introduction of approximate values of these quantities has only a slight effect upon the final value of the moment.

The deflexion is, of course, determined at the free edge in the usual way.

Numerical example

A right-angled wing wall with sides of 6, 8 and 10 m is free along the hypotenuse, simply supported along the long side enclosing the right-angle and restrained along the short side with i = 1. The wall is loaded with a triangular earth pressure, which is zero at the free edge and 2000 kg/m² at the right-angled corner. This gives $P = \frac{1}{3} \times 2000 \times 24 = 16000$ kg. By means of (3.23″), with i = 1 and 0, we get

$$m = \frac{3}{4} \frac{16000}{3\left(2\sqrt{2} + \frac{3}{4}\right)} = 1120 \text{ kg}$$

for the triangular loading. With two restraints, the correction factor is 1, and with simple supports, 1·20, in accordance with (3.271). With a single restraint, we put k = 1·10, which gives m = 1230 kg. We find a slab thickness of 13 cm, and the deflexion is found to be

$$u = \frac{1230 \times 1000^2}{8 \times 2 \times 10^5 \times \frac{1}{12} \times 13^3} = 4{\cdot}2 \text{ cm}$$

$$> \frac{1000}{250} = 4 \text{ cm}$$

The modified requirements to stiffness from 1.61 are applied. The slab thickness must thus be increased to 14 cm.

3.3 Isosceles triangle simply supported on one side and a column

The same observations apply to this as to the slabs in section 1.6.

Because of the symmetry, the yield line can be assumed to be parallel with the base a at a distance x from the free corner and thus has a length of (a/h)x. The two nodal forces shown act at the edges.

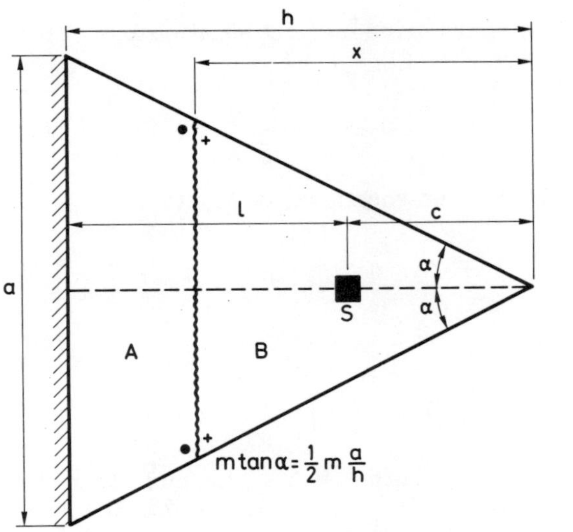

Fig. 3.3

The moment equation about a yields

$$m\frac{a}{h}x + m\frac{a}{h}(h - x) = \frac{1}{6}p(h - x)^2\left(a + 2\frac{a}{h}x\right)$$

or

$$m = \frac{1}{6}ph^2\left(1 - \frac{x}{h}\right)^2\left(1 + 2\frac{x}{h}\right) \tag{3.31}$$

The moment equation about an axis through the column parallel with a yields

$$m\frac{a}{h}x - m\frac{a}{h}(x - c) = \frac{1}{2}p\frac{a}{h}x^2\left(\frac{2}{3}x - c\right)$$

or

$$m\frac{c}{h} = \frac{1}{6}ph^2\left(\frac{x}{h}\right)^2\left(2\frac{x}{h} - 3\frac{c}{h}\right) \tag{3.32}$$

From these equations we find that

$$2\frac{x^3}{h^3} = \frac{c}{l} \tag{3.33}$$

from which x and thus also m can be found. Comparative calculations show that we have, to a good approximation,

$$m = \frac{1}{8}ph^2\frac{\frac{2}{3} - \frac{c}{h}}{\frac{1}{2} + \frac{c}{h}} \tag{3.34}$$

For $c = 0$, the correct formulae and the approximation give $m = \frac{1}{6}ph^2$ $(x = 0)$. When the column is located at the centroid of the triangle, we get $m = 0, x = h$.

The column reaction is statically determinate:

$$S = \frac{1}{6}ph^2\frac{a}{l} \tag{3.35}$$

The negative moment is $\frac{1}{6}pc^2$.

The conditions at the column are investigated by means of the formulae in YLT, pages 123-124.

3.4 Equilateral triangle supported on columns at the corners

The observations made in section 1.6 also apply to this type of slab.

With the yield pattern in Fig. 3.41, where all nodal forces are zero, the moment equation about the axis of rotation shown for the corresponding slab part yields

$$m\frac{a}{2} = \frac{P}{3}(x - c)$$

where x is the distance to the centroid, which is $\frac{1}{3}(\frac{2}{3}h + \frac{1}{2}h) = \frac{7}{18}h = \frac{7}{18}a(\sqrt{3}/2)$. Inserting this, we get

$$m = P\left(\frac{7}{18\sqrt{3}} - \frac{2}{3}\frac{c}{a}\right) = \frac{P}{\sqrt{3}}\left(\frac{7}{18} - \frac{c}{h}\right) \tag{3.41}$$

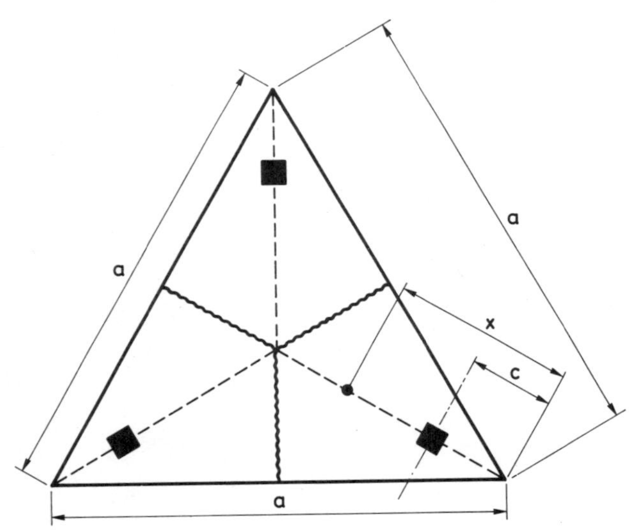

Fig. 3.41

For $c = 0$, we get $m = P/4{\cdot}45$.

There is, however, also a possibility that the yield pattern in Fig. 3.42 may occur. In this case, the nodal forces are $m \cot 60° = m/\sqrt{3}$.

Each of the small triangles has the load $P(x^2/h^2)$. For slab part A, the moment equations about the axes of symmetry are automatically satisfied. The equation of projection gives

$$6\frac{m}{\sqrt{3}} = P - 3P\frac{x^2}{h^2}$$

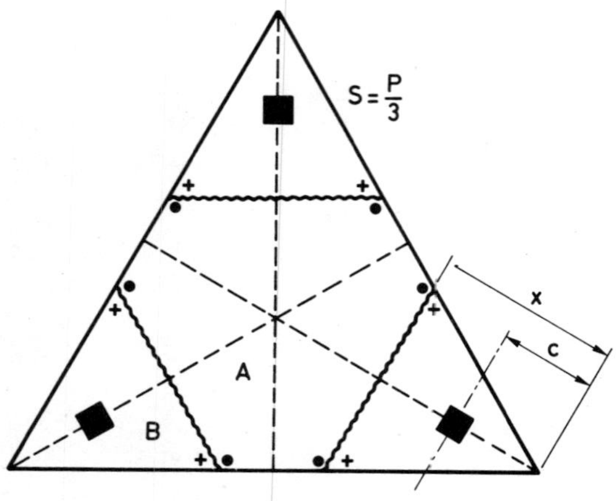

Fig. 3. 42

For slab part B, the moment equation about the yield line gives

$$m\frac{2x}{\sqrt{3}} = \frac{P}{3}(x-c) - P\frac{x^2}{h^2}\frac{x}{3}$$

From these two equations, we find that

$$\frac{x^3}{h^3} = \frac{c}{2h} \qquad (3.42)$$

and thereby that

$$m = \frac{P}{2\sqrt{3}}\left[1 - 3\left(\frac{c}{2h}\right)^{2/3}\right] \qquad (3.43)$$

For $c = 0$, we get $m = P/3{\cdot}47$, i.e. more than obtained from (3. 41). When $c/h = 0{\cdot}125$, the two yield patterns give the same value of m, so (3. 41) should only be used for higher values of c/h.

4 Circular slabs

Circular slabs are encountered in the form of bottom plates and covers for containers, wells and similar structures, and as footings for columns, foundation slabs, etc.

The load is frequently symmetrical about the centre, so we will confine ourselves to this case (a number of asymmetrical cases are dealt with in YLT, pages 23-25).

The worst, but still the most frequently used type of reinforcement is isotropic reinforcement, i.e. the same reinforcement in two directions at right-angles to each other. We will therefore begin with the treatment of the polar method of reinforcement, i.e. rings about the centre and straight bars along the radii. The rings may possibly be replaced by spirals whereby the number of laps can be reduced. At the centre we let two bars intersect at right-angles.

4.0 Polar reinforcement

We will begin by assuming that the reinforcement is variable, and so the bending moments m_φ in radial sections and m_r in circular sections are variable. Because of the symmetry, however, they are only functions of the radius r.

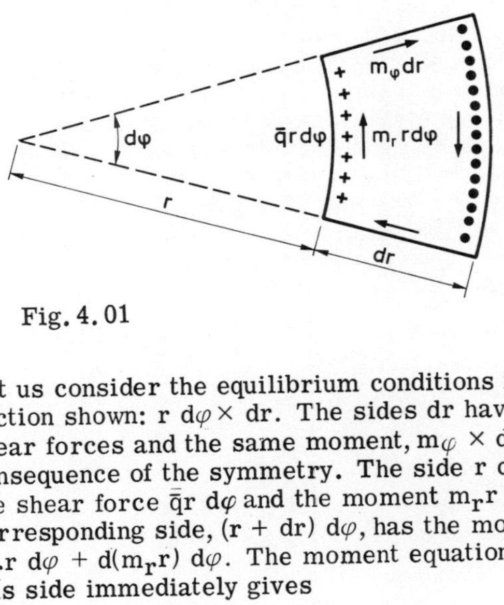

Fig. 4.01

Let us consider the equilibrium conditions for the section shown: $r\, d\varphi \times dr$. The sides dr have no shear forces and the same moment, $m_\varphi \times dr$, as a consequence of the symmetry. The side $r\, d\varphi$ has the shear force $\bar{q} r\, d\varphi$ and the moment $m_r r\, d\varphi$. The corresponding side, $(r + dr)\, d\varphi$, has the moment $m_r r\, d\varphi + d(m_r r)\, d\varphi$. The moment equation about this side immediately gives

$$m_r\, r\, d\varphi + m_\varphi\, dr\, d\varphi - m_r\, r\, d\varphi - d(m_r\, r)\, d\varphi = \bar{q}\, r\, d\varphi\, dr$$

since the moment due to the load is an infinitesimal of third order, and can therefore be neglected. Dividing by $d\varphi\, dr$, we get

$$m_\varphi - \frac{d(rm_r)}{dr} = \bar{q} r \qquad (4.01)$$

The equation of projection for the entire circle with radius r immediately gives

$$2\pi r \bar{q} = 2\pi \int_0^r pr\, dr = P_r$$

where P_r is the load within the circle mentioned. We thereby get

$$\bar{q}\, r = \frac{P_r}{2\pi} = \int_0^r pr\, dr \qquad (4.02)$$

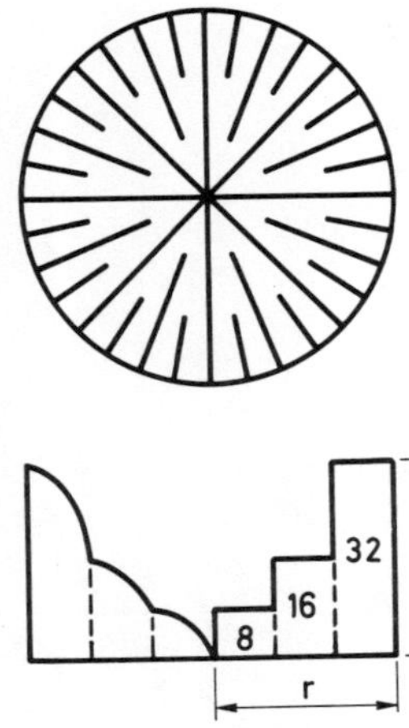

Fig. 4.02

The ring reinforcement corresponds to m_φ and the bar reinforcement to m_r. As long as the number of bars per circle remains unaltered, rm_r will remain constant. The slab shown in Fig. 4.02, in which there are 8, 16 and 32 bars per circle for the inner, intermediate and outer thirds of the slab, respectively, will, in theory, have the "stepped" curve for rm_r shown on the right. However, as the bars cannot transfer their force suddenly at the end cross-section, but only over a certain anchorage length, the actual course of rm_r will be as shown on the left in the Figure. It should be noted particularly that rm_r must be zero at the centre as m_r is finite.

If the bars were welded to a ring, the theoretical course could be realized. The ring should then be designed according to the tension in the bars. The corresponding compression at the top could then be imagined to be absorbed by a concrete compression ring at the top. The compression in this and the tension in the bottom ring correspond to a moment M_φ, which can incidentally be found by means of (4.01). Multiplying this by dr and integrating, we get

$$\int_1^2 m_\varphi dr = \int_1^2 \bar{q} r\, dr + (rm_r)_2 - (rm_r)_1 \qquad (4.03)$$

If we now let the limits of integration approach each other, the integral of the finite quantity $\bar{q}r$ will approach zero, while the difference corresponding to rm_r will assume the value $\triangle rm_r$ of the jump. m_φ must therefore tend towards infinity in such a way that its integral assumes the non-zero value:

$$M_\varphi = \lim_{1 \to 2} \int_1^2 m_\varphi \, dr = \triangle rm_r \qquad (4.04)$$

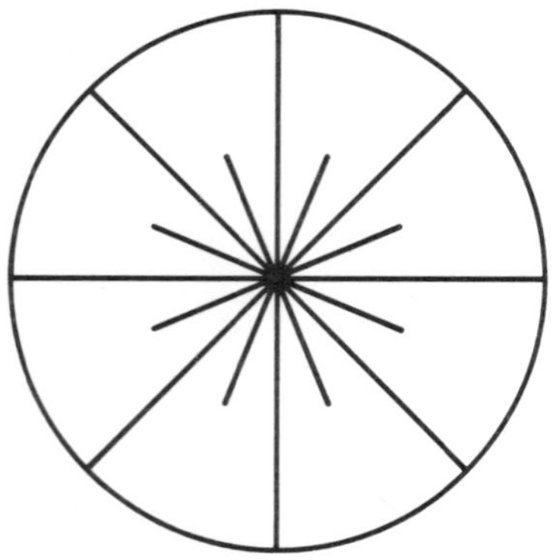

Fig. 4.03

If the reinforcement is arranged as shown in Fig. 4.03, the short bars can be anchored without welding if they are imagined to be anchored in a concrete compression ring by means of hooks. The compression at the top corresponding to the bars must, on the other hand, be absorbed by a tension ring. This means that M_φ is now negative. If there are reactions or a line load at the ends of the bars, the anchorage can be effected by means of these exactly as in beams (cf. Bygningsstatiske Meddelelser, 1945, No. 2). Whereas, in beams, the tensile bars must not end in the tensile zone inside the supports unless stirrups are arranged, this is possible in circular slabs if the ring corresponding to M_φ is provided.

If we assume the continuous variation of rm_r shown on the left in Fig. 4.02, the ring reinforcement will ensure the anchorage of the bar reinforcement.

The ring reinforcement can easily be completely adapted to a given moment curve, e.g. if the rings are to have the same dimension, the m_φ curve is divided into equal areas corresponding to the moment $\triangle m_\varphi$ that can be absorbed by one ring, and one ring is placed in each area (Fig. 4.04). The rm_r curve, not the m_r curve, is used for the bar reinforcement.

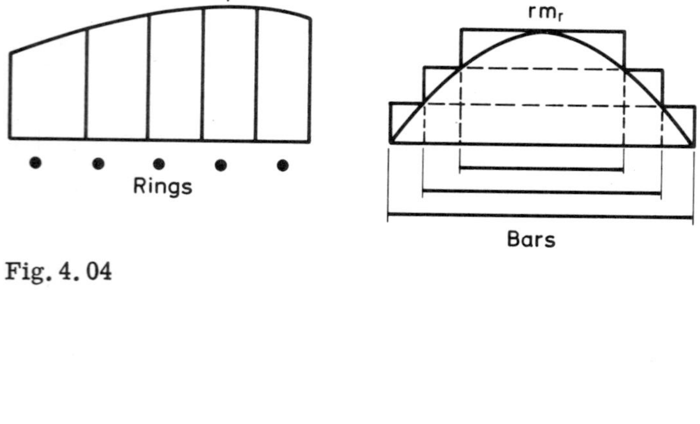

Fig. 4.04

The reinforcement can then be adapted to this by means of a stepped line, as shown. The circular yield lines will then only form where the stepped line touches the curve, i.e. at $\max(rm_r)$ and possibly at the jumps in the stepped line. As $\max(rm_r)$ does not usually coincide with $\max m_r$, we have here a case in which failure does not occur at maximum moment. Because of the special reinforcement, the maximum stress in the steel occurs at $\max(rm_r)$, and the maximum stress in the concrete, at $\max m_r$. However, we can safely design the slab for $\max(rm_r)$, even though we may thereby get a concrete stress at $\max m_r$ that is slightly in excess of the permissible value. The fact is that when we increase the load beyond the permissible value, until yielding begins at $\max(rm_r)$, the concrete stress at $\max m_r$ lies far from the ultimate value since, according to the assumptions of the yield-line theory, the slab is weakly reinforced. When the load is further increased up to failure, the concrete stress will not increase greatly at $\max m_r$, but only at $\max(rm_r)$, where the yielding results in wide cracks and reduced compressive zone, so that the concrete is finally crushed here.

We have only one equation for determining the moments m_φ and m_r, so we can select one of these and then determine the other. However, in this choice, due account must be taken of the geometrical conditions that determine the signs for the moments. As the surface of deflexion is a surface of rotation, the radius of curvature corresponding to m_r will be equal to that of the meridian curve, and the radius of curvature corresponding to m_φ will be equal to the length of the normal reckoned to the axis of rotation. Consequently, m_r changes its sign at the points of inflexion of the meridian curve, and m_φ at the points with horizontal tangent, i.e. at the points of maximum or minimum deflexion. As the bar reinforcement cannot be completely adapted to the rm_r curve, the meridian curve usually gets a corner at $\max(rm_r)$, where the yield line is formed. It follows from the foregoing remarks that m_φ changes its sign here. If m_φ is continuous, it must consequently be zero, and according to (4.01), $\bar{q}r$ will then also be zero since $d(rm_r) = 0$.

If only ring reinforcement is used, we get from (4.01)

$$m_r = 0, \; m_\varphi = \frac{P_r}{2\pi} \qquad (4.05)$$

which can only be used for simple support since the restraining moments correspond to m_r.

If only bars are used, we get, by differentiating (4.01) and (4.02),

$$m_\varphi = 0, \frac{d^2(rm_r)}{dr^2} = -pr \tag{4.06}$$

which shows that rm_r is determined as the moment in a beam with the load pr. This can also be seen directly by means of a sector element with $d\varphi = 1$.

In practice, we do not usually use either of these two special arrangements but a combination of both. It is therefore reasonable to seek the variation of m_φ that gives the smallest amount of reinforcement. If m_φ and m_r are both positive throughout the slab, the amount of reinforcement will be proportional to

$$\int_0^{2\pi} d\theta \int_a^b (m_\varphi + m_r)r\,dr = 2\pi \int_a^b \left(\bar{q}r^2 + r\frac{d(rm_r)}{dr} + rm_r\right) dr = 2\pi \int_a^b \bar{q}r^2\,dr + bm_b - am_a$$

where the integral depends solely upon the load and the last terms depend upon the restraining moments. If these are zero, the amount of reinforcement will not depend upon the choice of m_φ, and so the problem in this case is indeterminate. As only ring reinforcement can be completely adapted to the moment curve, the solution will, in practice, be $m_r = 0$.

When m_φ and m_r have opposite signs, we get the minimum amount of reinforcement for $m_\varphi = 0$, as will be demonstrated in chapter 6. In this case, (4.06) applies, i.e. the slab can be considered as divided into sector-shaped beams, so the true slab effect with beam action in two directions is no longer present. This means that the moments become very great in relation to the moments in the case of isotropic reinforcement. The saving in steel is thus obtained at the cost of the slab thickness and is therefore usually of no practical importance. In the case of very flexible slabs (4.21 and 4.22), the slab thickness may have to be increased so much with a view to the stiffness that the arrangement with weak ring reinforcement or none at all can be used.

In the case of stiff slabs, we arrange the reinforcement in such a way that the moments and thus also the thickness of the slab are the same as for isotropic reinforcement. This can usually be attained by making m_φ constant and equal to the value of m corresponding to isotropic reinforcement. The value of rm_r can then be determined by means of (4.01). We thereby usually achieve a not inconsiderable saving in steel in relation to isotropic reinforcement. When m_φ is constant, the rings must lie at constant intervals or they can be replaced by a spiral with constant pitch (Archimedes' spiral).

4.1 Simple circular slab

4.11 Isotropic reinforcement

An isotropic slab with symmetrical loading has a yield pattern with the radii as positive yield lines and, in case of restraint, a negative yield line at the support. Because of the symmetry, there are no shear forces along the radii of the sector shown in Fig. 4.111. The equation of projection gives the reaction as follows:

$$\bar{q}a\,d\varphi = \int_0^a pr\,dr\,d\varphi$$

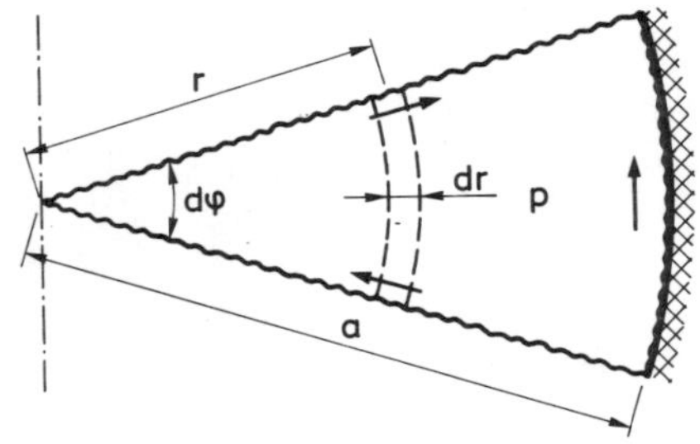

Fig. 4.111

The moment equation about the axis shown further gives

$$(m + m')a\,d\varphi = a\bar{q}a\,d\varphi - \int_0^a pr^2\,dr\,d\varphi$$

From these we get

$$m + m' = \int_0^a pr\left(1 - \frac{r}{a}\right)dr \tag{4.111}$$

Thence, for a uniformly distributed load p, a line load $\bar{p}$ acting along the circle with radius b, i.e.

$$\int p\,dr \rightarrow \bar{p}$$

and a point load P at the centre, i.e. $2\pi\bar{p}b \rightarrow P$ for $b \rightarrow 0$, we get

$$m + m' = \frac{1}{6}pa^2 + \bar{p}b\left(1 - \frac{b}{a}\right) + \frac{P}{2\pi} \tag{4.112}$$

For a conical load distribution that is p at the centre and zero at the edge, we get

$$m + m' = \int_0^a pr\left(1 - \frac{r}{a}\right)^2 dr = \frac{pa^2}{12} \tag{4.113}$$

The same value is obtained when the load is zero at the centre and p at the edge.

In restrained slabs, the top reinforcement must be carried in to the circle with radius ρ, which is determined by the following condition: if the slab part enclosed by the circle is calculated as a simply supported slab, the moment obtained must not exceed the value calculated previously, i.e.

$$m = \int_0^\rho pr\left(1 - \frac{r}{\rho}\right) dr \tag{4.114}$$

From this, we get the following.

For uniformly distributed load:

$$m = \frac{1}{6}\, p\rho^2, \quad \frac{\rho}{a} = \sqrt{\frac{m}{m + m'}}$$

For line load $\bar{p}$:

$$m = \bar{p}b\left(1 - \frac{b}{\rho}\right), \quad \frac{\rho}{a} = \frac{m + m'}{m + m'\,\frac{a}{b}}$$

For point load:

$$\rho = 0$$

For conical load:

$$\frac{\rho^2}{a^2}\left(1 - \frac{\rho}{a}\right) = \frac{m}{m + m'}$$

when p is zero at the edge, and

$$\frac{\rho}{a} = \sqrt[3]{\frac{m}{m + m'}}$$

when p is zero at the centre. When $m = m'$, we get $\rho/a = 0{\cdot}7$, $2b/(a + b)$, 0, $0{\cdot}6$ and $0{\cdot}8$ respectively.

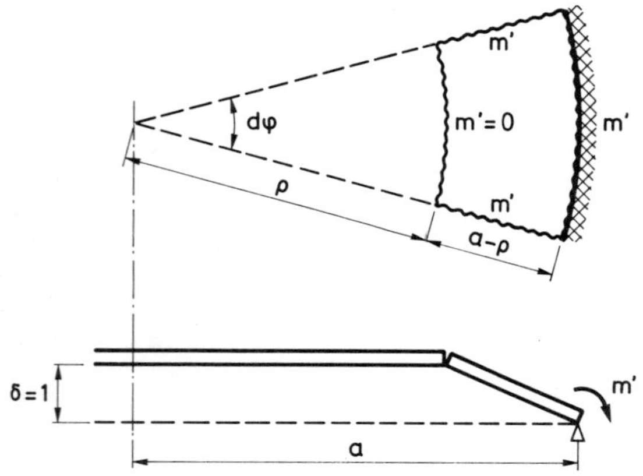

Fig. 4.112

If the restraint is such that m' may occur even though the slab is unloaded, as for example in the case of a cantilever slab, the values specified for ρ are not always sufficient. When the slab is subjected to m' and the dead load g, the yield pattern shown in Fig. 4.112 may be formed, in which the middle part is lifted δ and negative yield lines are formed along the limit of the top reinforcement and along the radii between this and the edge. For the sector shown, with $\delta = 1$, the work equation immediately gives

$$m'\rho\, d\varphi\, \frac{1}{a - \rho}$$

$$= \frac{1}{2}\, g\rho^2\, d\varphi \times 1 + \left(\frac{1}{6} ga\, d\varphi + \frac{1}{3}\, g\rho d\varphi\right)(a - \rho)^2\, \frac{1}{a - \rho}$$

or

$$m' = \frac{1}{6}\, g\, \frac{a^3 - \rho^3}{\rho} \tag{4.115}$$

from which ρ can be determined. We determine $m + m'$ by means of (4.111) or by means of the following, with $p + g$ instead of p. For uniformly distributed load, we get the special case of $m + m' = \frac{1}{6}(p + g)\, a^2$, which, using (4.115), gives

$$1 + \frac{p}{g} = \frac{m + m'}{m}\left(\frac{a}{\rho} - \frac{\rho^2}{a^2}\right) \tag{4.116}$$

The transition to this formula from that found earlier, $\rho/a = \sqrt{m/(m + m')}$, takes place when

$$\frac{p}{g} \geqslant \frac{m + m'}{m'}\left(\sqrt{\frac{m + m'}{m}} - 1\right) \tag{4.117}$$

In particular, $m = m'$ gives $p \geqslant 0{\cdot}83\, g$.

The deflexion is maximum at the centre and is determined by means of a strip along a diameter, which immediately gives

$$EIu = \frac{1}{2}\, ma^2 \tag{4.118}$$

In accordance with this, a simply supported slab with uniformly distributed load gets $EIu = pa^4/12$, whilst, according to the theory of elasticity, $EIu = pa^4/12{\cdot}8$.

4.12 Polar reinforcement

With $m_\varphi = m =$ constant, (4.01) immediately gives

$$rm_r = rm - \int_0^r \bar{q}r\, dr = rm - \int_0^r p\rho\,(r - \rho)\, d\rho \tag{4.121}$$

For $r = a$, the integral, according to (4.111), is equal to $a(m + m')$, i.e. $m_r = -m'$.

Corresponding to (4.112) and (4.113), we get, by means of (4.121),

$$rm_r = \begin{cases} rm - \dfrac{1}{6}\, pr^3 - \dfrac{P}{2\pi}\, r & \text{for } r \leqslant b \\ rm - \dfrac{1}{6}\, pr^3 - \bar{p}b\,(r - b) - \dfrac{P}{2\pi}\, r & \text{for } r \geqslant b \end{cases} \tag{4.122}$$

$$\left.\begin{aligned} rm_r &= rm - \frac{1}{12}\, pr^3\left(2 - \frac{r}{a}\right) \quad (p = 0 \text{ at the edge}) \\ rm_r &= rm - \frac{1}{12}\, p\, \frac{r^4}{a} \quad (p = 0 \text{ at the centre}) \end{aligned}\right\} \tag{4.123}$$

With $r = a$ and $m_r = -m'$, we again get from this (4.112) and (4.113). Substituting m' for m, we obtain

$$rm_r = \begin{cases} r\left[\dfrac{1}{6}\, p\,(a^2 - r^2) + \bar{p}b\left(1 - \dfrac{b}{a}\right)\right] - rm' & \text{for } r \leqslant b \\ \dfrac{1}{6}\, p\,(a^2 - r^2)\, r + \bar{p}b^2\left(1 - \dfrac{r}{a}\right) - rm' & \text{for } r \geqslant b. \end{cases} \tag{4.124}$$

$$\left.\begin{aligned} rm_r &= \frac{1}{12}\, pa^2r\left(1 - \frac{r^3}{a^3}\right) - rm' \\ rm_r &= \frac{1}{12}\, pa^2r\left(1 - \frac{r}{a}\right)\left(1 + \frac{r}{a} - \frac{r^2}{a^2}\right) - rm' \end{aligned}\right\} \tag{4.125}$$

The bar reinforcement is determined from these formulae.

For simple support, $m' = 0$, we get only positive values of rm_r, and the maximum value is decisive. For a point load only, we get $rm_r = 0$. If there is also a uniformly distributed load, we get max $(rm_r) = pa^3/9\sqrt{3} = pa^3/15{\cdot}6$. When there is a line load as well, the maximum value will normally occur for $r = b$, which gives

$$\max (rm_r) = \frac{1}{6}\, pa^2 b\left(1 - \frac{b^2}{a^2}\right) + \bar{p}b^2\left(1 - \frac{b}{a}\right)$$

It is on the safe side to assume that

$$\max (rm_r) = \frac{pa^3}{15{\cdot}6} + \bar{p}b^2\left(1 - \frac{b}{a}\right) \qquad (4.126)$$

For a conical load distribution we get

$$\left.\begin{aligned} \max (rm_r) &= \frac{5pa^3}{192} \text{ for } r = \frac{a}{2} \\ \frac{pa^3}{16\,\sqrt[3]{4}} &= \frac{pa^3}{25{\cdot}4} \text{ for } r^3 = \frac{a^3}{4} \end{aligned}\right\} \qquad (4.127)$$

In the case of restraint, rm_r assumes the value zero at a point determined by

$$m = \int_0^r p\rho\left(1 - \frac{\rho}{r}\right)d\rho$$

which corresponds exactly to (4.114) when the symbols ρ and r are interchanged. This means that rm_r assumes the value zero exactly at the limit of the top reinforcement in the corresponding isotropic slab. In other words, isotropic reinforcement and polar reinforcement give the same limit for the top reinforcement at the restrained edges.

The slab part enclosed by these limits can be considered as simply supported and be reinforced according to the foregoing formulae. The negative values of rm_r are maximum at the restraint, and here equal to $-am'$. For uniformly distributed load we can, once and for all, plot the curve $y = x(1 - x^2)$, which, together with the straight line $y = (6m'/pa^2)x$ corresponding to m', determines $rm_r = \frac{1}{6}\, pa^3 y$ for $x = r/a$. The location of the ends of the bar reinforcement is then determined by means of stepped lines parallel with the straight line.

Finally, we will compare the amount of steel required for slabs of equal thickness, subjected to a uniformly distributed load, with isotropic reinforcement and polar reinforcement respectively. As the amount of steel is proportional to

$$\int_0^a (|\, m_\varphi \,| + |\, m_r \,|) r\, dr = J$$

simple support and isotropic reinforcement, i.e. $m_\varphi = m_r = m = \frac{1}{6}\, pa^2$, give $J_i = \frac{1}{6}\, pa^4$. In the case of polar reinforcement, with $m_\varphi = m = \frac{1}{6}\, pa^2$ and $rm_r = \max (rm_r) = pa^3/15{\cdot}6$, we get

$$J_p = \frac{pa^4}{12} + \frac{pa^4}{15{\cdot}6} = \frac{pa^4}{6{\cdot}8}$$

If rm_r is stepped down to half, which can be done at $r = 0{\cdot}2a$ and $r = 0{\cdot}89a$ (Fig. 4.113), we get $\frac{1}{2}(0{\cdot}2 + 0{\cdot}11)a\ (pa^3/15{\cdot}6)$ less, i.e. $J_p = pa^4/7{\cdot}3$. Isotropic reinforcement thus requires 13% more steel in the first case and 22% more in the second.

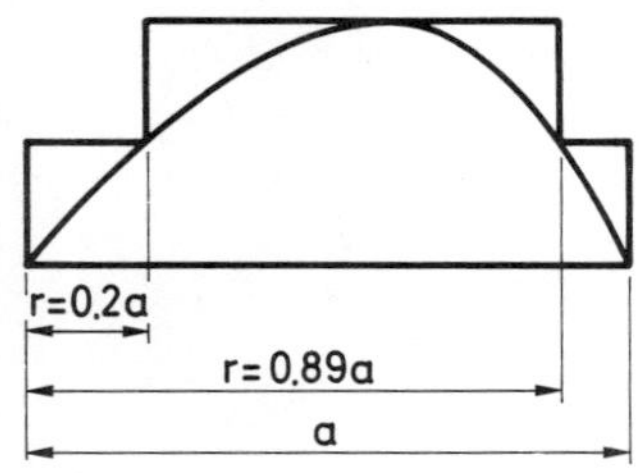

Fig. 4.113

In the case of restraint, with $m' = m$, isotropic reinforcement gives $m_\varphi = m_r = pa^2/12$ over the whole slab and $m'_\varphi = m'_r = pa^2/12$ from $r = a/\sqrt{2}$ to $r = a$. We thus get

$$J_i = \frac{pa^2}{12}\left(a^2 + \frac{1}{2}\, a^2\right) = \frac{pa^4}{8}$$

For polar reinforcement, with $m_\varphi = m = pa^2/12$, rm_r changes signs at $r = a/\sqrt{2}$. The part lying within this, which can, as mentioned, be considered as simply supported, contributes $pa^4/(4 \times 6{\cdot}8)$, and the remainder contributes $\frac{1}{2}\,(pa^2/12)\,(a^2/2)$ from m_φ and $(pa^3/12) \times (1 - \sqrt{\frac{1}{2}})$ a from rm_r = constant $= -am'$, making a total of $J_p = pa^4/12{\cdot}2$. Thus, in this case, isotropic reinforcement requires 53% more steel.

As isotropic reinforcement obviously entails greater waste, it actually requires even more steel.

4.2 Ring-shaped slabs

If the slab is only supported around one circle, the ring inside and outside this can be calculated by means of the formulae in 4.21 and 4.22.

4.21 Supported along the inner edge, free at the outer edge

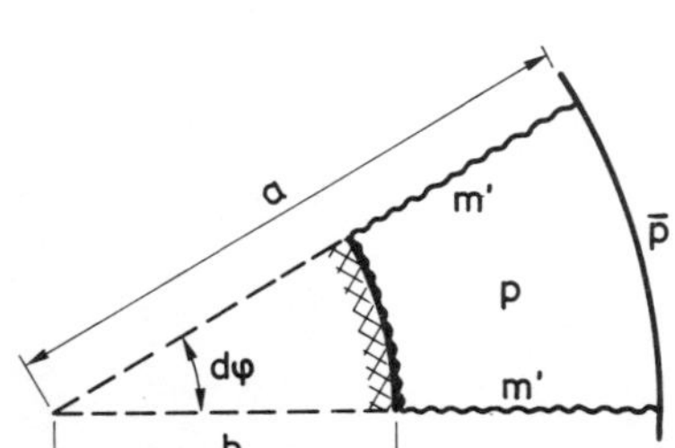

Fig. 4.211

We will first consider the isotropically reinforced

slab. The slab is only provided with top reinforcement. The moment at the restraint is $m_b' = im'$ where m' is the moment in the slab. In the case of simple support, we have $i = 0$, and in the case of restraint, $i = 1$. With the loads indicated, the moment equation about the support for the sector shown gives

$$im'b\, d\varphi + m'(a - b)\, d\varphi = \frac{1}{6} p\, (a - b)^2 b\, d\varphi$$

$$+ \frac{1}{3} p\, (a - b)\, a\, d\varphi + \bar{p}a\, (a - b)\, d\varphi$$

or

$$m'[a - (1 - i)b] = \frac{1}{6} p\, (a - b)^2\, (2a + b) + \bar{p}a\, (a - b) \qquad (4.211)$$

For $i = 0$ we get

$$m' = \frac{1}{6} p\, (a - b)\, (2a + b) + \bar{p}a \qquad (4.212)$$

and for $i = 1$

$$m' = \frac{1}{6} p\, (a - b)^2 \left(2 + \frac{b}{a}\right) + \bar{p}\, (a - b) \qquad (4.212')$$

The deflexion is found by means of the strip shown in Fig. 4.212, which gives

$$EIu = \frac{1}{a}\, m'\, (a^2 - b^2) \times 1 = 2\sqrt{a^2 - b^2} \qquad (4.213)$$

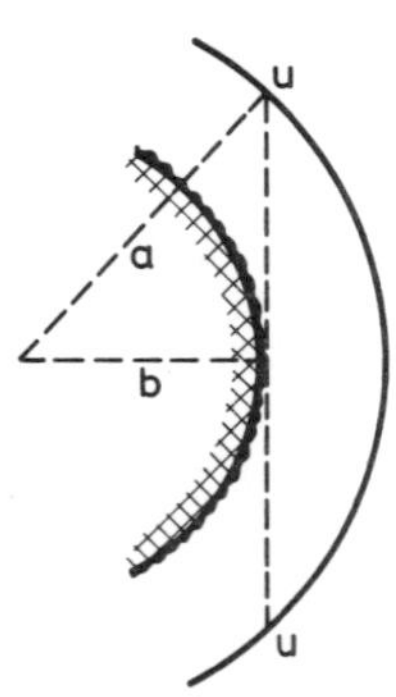

Fig. 4.212

In the case of polar reinforcement, we get the following from (4.01), with $\bar{q}r = \frac{1}{2}\, p\, (a^2 - r^2) - \bar{p}a$ and $m_\varphi = -m' =$ constant,

$$- rm_r = rm_r' = \frac{1}{6} p\, (a - r)^2 (2a + r) + \bar{p}a\, (a - r)$$

$$- m'\, (a - r) \qquad (4.214)$$

For $r = b$ and $m_r' = im'$, we again get (4.211).

With this value of m', we then find

$$rm_r'\, (a - b + ib) = \frac{1}{6} p\, (a - r)\, [ib\, (a - r)\, (2a + r)$$

$$- (a - b)\, (r - b)\, (a + b + r)] + i\bar{p}b\, (a - r) \qquad (4.215)$$

In the special cases, $i = 0$ and $i = 1$, we find

$$rm_r = \frac{1}{6} p\, (a - r)\, (r - b)\, (a + b + r)$$

with m_r positive (4.216)

$$rm_r' = \left[\frac{1}{6} p \left(3ab - \frac{b^3}{a} - ar - r^2\right) + \bar{p}b\right] (a - r) \qquad (4.216')$$

For simple support, we get a very nearly parabolic variation of rm_r, and the maximum value can, to a good approximation, be found by putting $r = (a + b)/2$,

$$\max(rm_r) \sim \frac{p}{16}\, (a - b)^2 (a + b) \qquad (4.217)$$

In the case of restraint, there will usually be a point at which $rm_r = 0$. Outside this, the conditions are as in the simply supported ring, and max (rm_r) can be found by means of (4.217) when b is replaced by the radius to the point of zero moment. Inside, rm_r' increases to the restraint value bm', which is the maximum value for the top reinforcement.

By comparison with (4.212), (4.217) usually gives only a very light bar reinforcement. As this reinforcement has to lie at the bottom, and the rings or spirals at the top, it would be convenient simply to do without it. We then get, for simple support,

$$m_\varphi' = - m_\varphi = - \bar{q}r = \frac{1}{2} p(a^2 - r^2) + \bar{p}a \qquad (4.218)$$

with maximum value $\frac{1}{2}\, p\, (a^2 - b^2) + \bar{p}a$.

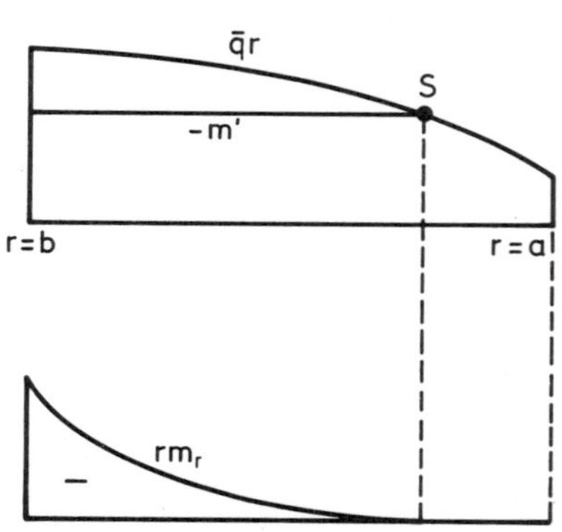

Fig. 4.213

The light bar reinforcement at the bottom of the outermost part is similarly inconvenient in the case of the restrained slab. It can be omitted by making $m_\varphi = \bar{q}r$ outside the point of intersection S between the horizontal line corresponding to $m_\varphi = - m'$ = constant and the $\bar{q}r$ curve (Fig. 4.213). The derivative of rm_r thereby becomes zero over this length, and rm_r has only negative values.

If the ring reinforcement is omitted altogether, (4.214), with $m' = 0$, gives

$$rm_r' = \frac{1}{6} p(a - r)^2 (2a + r) + \bar{p}a\, (a - r) \qquad (4.219)$$

the maximum value of which occurs at the restraint $r = b$ and gives $m_b' = a/b$ times the value found from (4.212′).

Formulae (4.218) and (4.219) give much thicker slabs than the foregoing formulae, but as these slabs are very flexible, they must still be made considerably thicker because of the stiffness requirements, so the latter formulae may well be used.

Numerical example

A circular tower with a radius of 4 m has a 1·5 m wide outside gallery, which has to be calculated for a load of 500 kg/m². If the tower is a brick structure, simple support will be the most obvious solution, and if it is of reinforced concrete, restraint can easily be provided. We will investigate both possibilities.

(1) Using (4.212) and with the total load 500 + 400 = 900 kg/m², simple support gives $m' = 1/6 \times 900 \times 1{\cdot}5 \times 15 = 3375$ kg, $h = 18$ cm. The deflexion is then

$$u = \frac{1}{2}\,\frac{3375(5{\cdot}5^2 - 4{\cdot}0^2)10^4}{2 \times 10^5 \times \frac{1}{12} \times 18^3}$$

$$= 2{\cdot}5 \text{ cm} > \frac{2\sqrt{5{\cdot}5^2 - 4{\cdot}0^2}}{500} \times 100 = 1{\cdot}5 \text{ cm}$$

The thickness of the slab must be increased to 22 cm, which gives a total load of 1030 kg/m² and $m' = 3860$ kg, $f = 18{\cdot}5$ cm² per m. With polar reinforcement, we get 16 mm dia. rings spaced at 11 cm. At the bottom, bars are provided corresponding to (4.217): $(1030/16) \times 1{\cdot}5^2 \times 9{\cdot}5 = 1375$ kgm. From this, we find that the moment at the wall is 1375/4 = 344 kg, which gives $f = 1{\cdot}7$ cm² per m, 7 mm dia. bars spaced at 23 cm.

With ring reinforcement alone, we use (4.218) to find the maximum value,

$$m' = 1/2 \times 1030\ (5{\cdot}5^2 - 4{\cdot}0^2) = 7350 \text{ kg}$$

which requires $h = 26$ cm, although only at the wall.

(2) Using (4.212′), we find that restraint with $i = 1$ gives $m' = 1/6 \times 900 \times 1{\cdot}5^2 \times 2{\cdot}73 = 921$ kg, $h = 11$ cm. The deflexion is

$$u = \frac{1}{2}\,\frac{921(5{\cdot}5^2 - 4{\cdot}0^2)10^4}{2 \times 10^5 \times \frac{1}{12} \times 11^3} = 3{\cdot}0 \text{ cm} > 1{\cdot}5 \text{ cm}$$

The thickness has to be increased to 14 cm. The load is then 836 kg/m², $m' = 856$ kg, $f = 7{\cdot}0$ cm² per m. If only bar reinforcement is used, (4.219) gives maximum $m' = 856 \times 5{\cdot}5/4 = 1180$ kg, which does not require increased thickness, but $f = 7{\cdot}0 \times 5{\cdot}5/4 = 9{\cdot}6$ cm², e.g. 10 mm dia. bars spaced at 8 cm.

For practical reasons, a certain amount of ring reinforcement is necessary. If, for example, we take 7 mm dia. rings spaced at 20 cm, this corresponds to $m_\varphi' = 856 \times 1{\cdot}92/7{\cdot}0 = 235$ kg $= m'$.

From (4.214) we then get $m_b' = 1180 - 235 \times 1{\cdot}5/4 = 1090$ kg, $f = 8{\cdot}55$ cm², i.e. 10 mm dia. bars spaced at 9 cm. The bar reinforcement at the bottom becomes so small that we can neglect it.

4.22 *Supported along the outer edge and free at the inner edge*

We get positive yield lines along the radii and negative yield lines around the circumference. In the case of isotropic reinforcement, the moment equation for the sector shown now gives

$$m'a + m(a - b) = \frac{1}{6}\,p(a - b)^2(a + 2b) + \bar{p}b(a - b) \qquad (4.221)$$

$m' = 0$ and $m' = m$ give

$$m = \frac{1}{6}\,p(a - b)(a + 2b) + \bar{p}b \qquad (4.222)$$

$$m = m' = \frac{1}{6}\,p(a - b)^2\,\frac{a + 2b}{2a - b} + \bar{p}b\,\frac{a - b}{2a - b} \qquad (4.223)$$

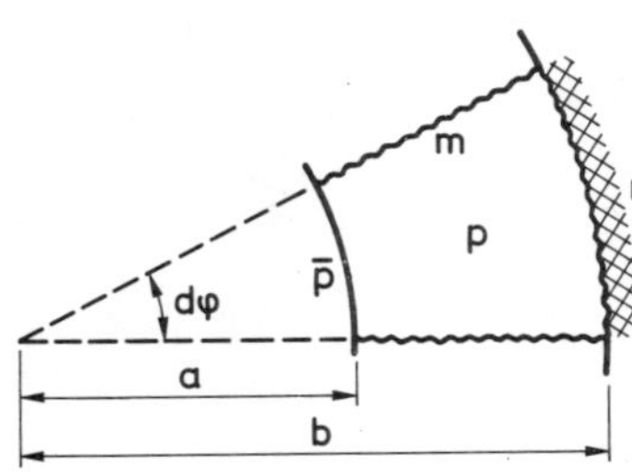

Fig. 4.22

The former corresponds exactly to (4.212). The deflexion is found in the same way and by means of the same formula as in section 4.21, with m instead of m′. In the case of polar reinforcement, with $m_\varphi = m$, $\bar{q}r = 1/2\,p(r^2 - b^2) + \bar{p}b$, we get

$$rm_r' = \frac{1}{6}\,p(r - b)^2(r + 2b) + \bar{p}b(r - b) - m(r - b)$$

$$= -\,rm_r \qquad (4.224)$$

which gives (4.221) for $r = a$ and $m_r' = m'$. Corresponding to (4.222), we get (4.225) equal to (4.216), and corresponding to (4.223), we get

$$(2a - b)rm_r' = \frac{1}{6}\,(r - b)[r(r + b)(2a - b)$$

$$- a(a^2 + b^2)] + \bar{p}ab(r - b) \qquad (4.226)$$

These formulae are used as in section 4.21.

Slabs with a hole at the centre are calculated in accordance with these formulae. Eccentric holes give asymmetrical yield patterns with complicated calculations. For small holes, we may use an approximation similar to that given in section 1.15 in connexion with reduced degree of restraint. For a small hole at the centre, which is assumed to be loaded in the same way as the slab, we now get $m'a + m(a - b) = m(1 + i - b/a)a$ instead of $(m + m')a = m(1 + i)a$, on account of the hole. A central hole thus reduces the degree of restraint by b/a. A slight eccentricity will naturally not make much difference to this. A hole with its centre on the periphery will approximately reduce the contribution from the moments to $(2\pi a - 2b)(m + m')$; i.e. the reduction in the degree of restraint is $\Delta i = (1 + i)(b/\pi a)$, which will not exceed b/a as long as $i \leq 2{\cdot}1$. Denoting the eccentricity of the hole c, we can interpolate parabolically between these values by means of

$$\Delta i = \left(1 - \frac{2 - i}{3}\frac{c^2}{a^2}\right)\frac{b}{a} \qquad (4.227)$$

where, for the sake of simplicity, π is approximated by 3.

4.23 Supported along both edges

Taking an isotropic slab with uniformly distributed load, we will formulate the work equation for the element shown in Fig. 4.231 with $d\varphi = 1$.

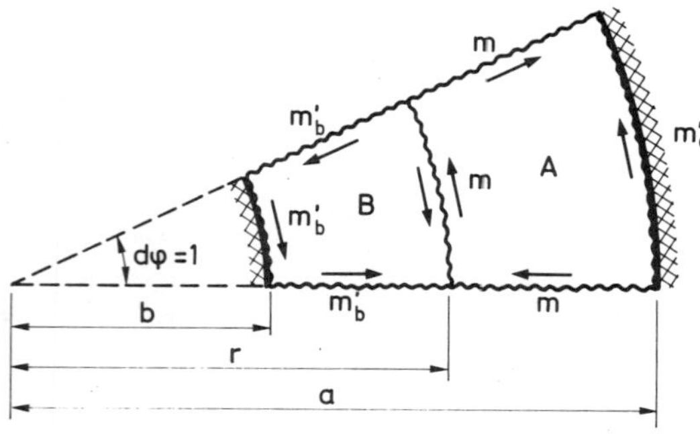

Fig. 4.231

The yield pattern is formed by a positive yield line with radius r between the supports and radial yield lines that are negative in slab part B and positive in slab part A. This follows from the observation in 4.0 regarding the determination of the signs for m_r and m_φ when considering the deformations. We find the maximum deflexion at the positive circular yield line, so m_φ changes sign here. We get negative yield lines along the supports in the case of restraint. We will assume that the negative yield lines in slab part B all have the same moment m'_b. The top reinforcement at the restrained edge must thus be taken right out and past the positive yield line. As the rotations can be put at $1/(a - r)$ and $1/(r - b)$, the contribution to the work equation from the element under consideration is

$$(m + m'_a)\frac{a}{a - r} + (m + m'_b)\frac{r}{r - b}$$

$$= \frac{1}{6}p(a - r)^2\frac{a + 2r}{a - r} + \frac{1}{6}p(r - b)^2\frac{b + 2r}{r - b}$$

$$= \frac{1}{6}p(a - b)(a + b + r) \qquad (4.231)$$

We now estimate that $r = \sqrt{ab}$, which gives the simple approximate formula

$$m'_a + 2m + m'_b = \frac{1}{6}pa(a - b)\left(1 - \frac{b}{a}\sqrt{\frac{b}{a}}\right) \qquad (4.232)$$

The exact solution is obtained by differentiating (4.231) with respect to r. We thereby get a fourth degree equation for the determination of r and thereby of the moments. These calculations give the following approximate formula:

$$m'_a + 2m + m'_b = \frac{p(a - b)^2}{15}\left(2{\cdot}75 + \frac{b}{a}\right) \qquad (4.233)$$

which gives values that are a few per cent higher than those from (4.232). By using the same calculations, the following approximate formulae were found for the reactions $\bar{q}_a$ and $\bar{q}_b$ at the outer and inner supports:

$$\bar{q}_a = p\frac{a - b}{4a}\,\frac{(2a + b)m'_a + (3a + b)m + am'_b}{m'_a + 2m + m'_b}$$

$$(4.234)$$

$$\bar{q}_b = p\frac{a - b}{4b}\,\frac{bm'_a + (a + 3b)m + (a + 2b)m'_b}{m'_a + 2m + m'_b}$$

The limits for the inner and outer top reinforcement both lie between the positive yield line and the outer edge, i.e. so close together that we would in practice provide top reinforcement over the whole slab. If only the outer edge is restrained, however, we can make do with carrying the top reinforcement at this edge a distance of $\frac{1}{5}(a - b)$ into the slab, provided the degree of restraint does not exceed 1.

As regards the deflexion, a strip along a radius gives $EIu = \frac{1}{8}m(a - b)^2$. In the case of polar reinforcement we have

$$\bar{q}r = \frac{1}{2}p(r^2 - \rho^2) \qquad (4.235)$$

where ρ is an unknown quantity. Further, we saw previously that m_φ is negative at the inner edge and positive at the outer edge. m_φ changes sign at the circular yield line, i.e. where rm_r is maximum. If m_φ varies continuously, it will consequently be zero where $d(rm_r)$ is zero, but then $\bar{q}r$ is also zero in accordance with (4.01). As this type of slab is not particularly flexible, we will seek a polar reinforcement that gives a slab thickness that is as close as possible to that obtained with isotropic reinforcement. If we now take $m_\varphi = -m'$ and $m_\varphi = m$ in the two slab parts, we get the conditions shown in Fig. 4.232, where the hatched curve represents $m_\varphi - \bar{q}r = d(rm_r)/dr$. We then get the variation of rm_r shown in Fig. 4.232.

This shows that the maximum value of rm_r does not occur at $r = \rho$, as assumed, so m_φ cannot be selected in this way.

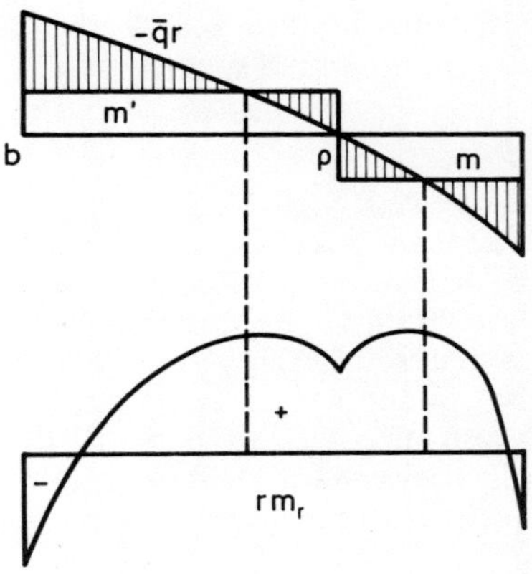

Fig. 4.232

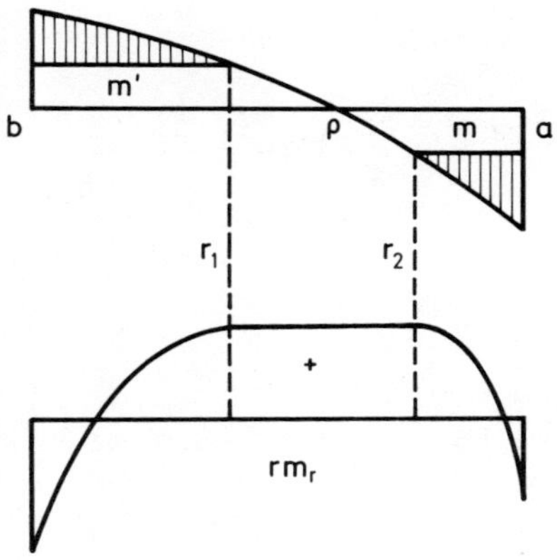

Fig. 4.233

If, on the other hand, m_φ is taken as constant until it intersects the $\bar{q}r$ curve at r_1 and r_2 and we select $m_\varphi = \bar{q}r$ between r_1 and r_2, whereby $d(rm_r) = 0$ here, we get the conditions shown in Fig. 4.233, where maximum rm_r now occurs over the entire length between r_1 and r_2. Our assumption will then be fulfilled. It will be seen that this choice of m_φ entails slightly greater moments than in the case of isotropic reinforcement.

We immediately get from (4.235)

$$\left.\begin{aligned} m' &= \frac{1}{2}\,p\,(\rho^2 - r_1{}^2),\; m = \frac{1}{2}\,p(r_2{}^2 - \rho^2) \\ &\text{or} \\ r_1{}^2 &= \rho^2 - \frac{2m'}{p},\; r_2{}^2 = \rho^2 + \frac{2m}{p} \end{aligned}\right\} \qquad (4.235')$$

By means of (4.01) and the boundary conditions, we further find that:

$$\left.\begin{aligned} &\text{for } r < r_1: \\ &rm_r = \frac{1}{2}\,pr_1{}^2(r-b) - \frac{1}{6}\,p(r^3 - b^3) - bm_b' \\ &\text{and} \\ &\text{for } r > r_2: \\ &rm_r = \frac{1}{2}\,pr_2{}^2(r-a) - \frac{1}{6}\,p(r^3 - a^3) - am_a' \end{aligned}\right\} \qquad (4.236)$$

The maximum value is obtained for $r = r_1$ and $r = r_2$, respectively, and must be the same in both cases. This gives one further equation which, together with the two in (4.235′), can be used to determine ρ, r_1 and r_2. Inserting m and m′ from (4.235′), we find that

$$\begin{aligned} \max(rm_r) &= \frac{1}{6}\,p(r_1 - b)^2(b + 2r_1) - bm_b' \\ &= \frac{1}{6}\,p(a - r_2)^2(a + 2r_2) - am_a' \end{aligned} \qquad (4.237a)$$

$$m + m' = \frac{1}{2}\,p(r_2{}^2 - r_1{}^2) \qquad (4.237b)$$

The latter equation is obtained by addition of the equations in (4.235′). Maximum m_r in the yield zone, which extends from r_1 to r_2, is obtained at r_1, and we should therefore have $\max(rm_r) = r_1m$. Furthermore, m_a' and m_b' must not exceed m′, and, in particular, we can assume $m' = m_b'$, as in the case of isotropic reinforcement. Introducing these relations and $m_a' = i_am$, $m_b' = i_bm$, we get the following equations for the determination of r_1 and r_2:

$$\begin{aligned} 3(r_1 + i_bb)(r_2{}^2 - r_1{}^2) &= (1 + i_b)(r_1 - b)^2(2r_1 + b) \\ 3(r_1 + i_aa)(r_2{}^2 - r_1{}^2) &= (1 + i_b)(a - r_2)^2(2r_2 + a) \end{aligned} \qquad (4.238)$$

When these have been determined, we find the moments by means of (4.237), ρ by means of (4.235′), and the reactions by means of (4.235), with $r = a$ and b. The numerical values are

$$a\bar{q}_a = \frac{1}{2}\,p(a^2 - \rho^2),\; b\bar{q}_b = \frac{1}{2}\,p(\rho^2 - b^2) \qquad (4.239)$$

By selecting r_1 and r_2, we can find the corresponding values of a and b comparatively easily. By doing this for the various restraint alternatives, we find, by means of comparative calculations, the following approximate formula corresponding to (4.233):

$$m_a' + 2m + m_b' = \frac{p(a-b)^2}{50}\left[9{\cdot}5 + 3\,\frac{b}{a} + (i_b - i_a)\left(1 - \frac{b}{a}\right)\right] \qquad (4.233')$$

It is further found that the point corresponding to ρ divides $a - b$ in the ratio $\sqrt{1 + i_a}/\sqrt{1 + i_b}$, i.e.

$$\rho = \frac{a\sqrt{1 + i_b} + b\sqrt{1 + i_a}}{\sqrt{1 + i_a} + \sqrt{1 + i_b}}$$

which gives the following reactions:

$$a\bar{q}_a = \frac{1}{2} p(a - b) \sqrt{1 + i_a} \frac{2a\sqrt{1 + i_b} + (a + b)\sqrt{1 + i_a}}{(\sqrt{1 + i_a} + \sqrt{1 + i_b})^2}$$

$$b\bar{q}_b = \frac{1}{2} p(a - b) \sqrt{1 + i_b} \frac{2b\sqrt{1 + i_a} + (a + b)\sqrt{1 + i_b}}{(\sqrt{1 + i_a} + \sqrt{1 + i_b})^2} \quad (4.234')$$

For $i_a = i_b$, we get the same expressions as in (4.234), whilst for $i_a \neq i_b$, the difference is only slight, so (4.234) can be used, i.e. the reactions are for all practical purposes the same as in isotropic slabs.

With the above-mentioned value of ρ and the moments determined by means of (4.233′), r_1 and r_2 can, if desired, be determined by means of (4.235′) and the variation of the moment by means of (4.236).

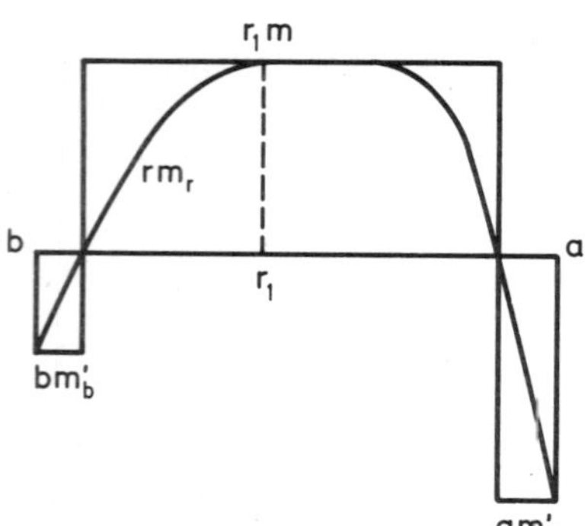

Fig. 4.234

As we would usually make do with simple bar reinforcement that corresponds to the constant values am_a' and bm_b' for rm_r at the restraints and r_1m at the middle, it will suffice to determine the limits for the top and bottom reinforcement, i.e. the points at which $rm_r = 0$ (Fig. 4.234). These are found by means of (4.236), which has the form $A = r(B - r^2)$, which is easily solved by using a slide rule. Here, we have

$$A = \begin{cases} b\left(6\dfrac{m_b'}{p} + 3r_1^2 - b^2\right) \\ a\left(6\dfrac{m_a'}{p} + 3r_2^2 - a^2\right) \end{cases} \quad (4.236')$$

$$B = \begin{cases} 3r_1^2 \\ 3r_2^2 \end{cases}$$

at the outer and inner support, respectively.

As a rule we would carry some of the innermost reinforcement right out to the supports. The top reinforcement can be carried a distance of $\frac{1}{5}(a - b)$ into the slab, provided the degree of restraint does not exceed 1. The exact calculations show that this is on the safe side.

Between r_1 and r_2, the ring reinforcement must be lighter (Fig. 4.233). Assuming a linear variation, we see that the mean values are $\frac{1}{2}$ m′ and $\frac{1}{2}$ m for the negative and positive moments, respectively.

Numerical example

A 300 m^3 water tank is designed as a circular tank with a diameter of 8 m and a depth of 6 m. The bottom consists of a slab supported along the wall of the tank and on a column at the middle. The slab is considered as restrained in the wall and in the capital of the column, which has a diameter of 2 m. With a 20 cm thick bottom slab, the load will be 6480 kg/m^2. With the degrees of restraint $i_a = i_b = 1$, we find from (4.233′), with a = 4·0 m and b = 1·0, that m = (6480/200) (10·25 × 9) = 3000 kg. With the stresses 60 and 1000 kg/cm^2, we then get $h_a = 0{\cdot}289\sqrt{3000} = 15{\cdot}8$ cm, $f = 0{\cdot}410\sqrt{3000} = 22{\cdot}4$ cm^2. With h = 20 cm and 20 mm dia. bars, $h_n' = 16{\cdot}0$ cm, and f′ = 22·2 cm^2 per m, i.e. 20 mm dia. bars spaced at 14 cm.

As $\rho = \frac{1}{2}(a + b) = 2{\cdot}5$, (4.239) gives $4\bar{q}_a = \frac{1}{2} \times 6480\,(4^2 - 2{\cdot}5^2)$, $\bar{q}_a = 7900$ kg/m, and $1 \times \bar{q}_b = \frac{1}{2} \times 6480/(2{\cdot}5^2 - 1^2) = 17000$ kg/m. The shear stress is $\tau = 17000/(0{\cdot}9 \times 16 \times 100) = 11{\cdot}8$ kg/cm^2. Inclined bars have to be provided at the column capital. At the wall, $\tau = 5{\cdot}5$ kg/cm^2, so inclined bars can be omitted here. We find from (4.235′) that $r_1 = 2{\cdot}31$ and $r_2 = 2{\cdot}68$, in metres.

We now get the following reinforcement.

Ring reinforcement at the top from b to r_1, i.e. 131 cm, calls for 131/14 = 9·35 rings, and from r_1 to ρ, 19 cm, rings at intervals of 28 cm, i.e. 19/28 = 0·68 rings, i.e. a total of ten 20 mm dia. rings.

Correspondingly, the ring reinforcement at the bottom calls for 18/28 + 132/14, i.e. also ten 20 mm dia. rings. The bar reinforcement at the top at both restraints will be 20 mm dia. bars spaced at 14 cm, which gives $2\pi \times 400/14 = 180$ bars and $2\pi \times 100/14 = 45$ bars at the outer and inner edges, respectively. Every fourth bar at the outer edge can be taken in to the inner edge, while the remainder can end or be bent down 60 cm from the edge.

The bar reinforcement at the bottom must be spaced at intervals of 14 cm at radius $r_1 = 231$ cm. This gives $2\pi \times 231/14 = 104$ bars, which can be taken out to the edges. At the inner edge, however, alternate bars can be bent up to resist the shear force.

4.3 Slabs on soil

These types of slab are encountered as roadways and as airport runways and taxiways. Obviously, such slabs are never circular, but they are designed for a point load, so the stress field is symmetrical about the load and can therefore be dealt with as the field in a circular slab with symmetrical loading.

4.31 Plastic soil

The soil is assumed to be of such a character that its reaction does not increase when the deformations exceed a certain value but remains constant at the plastic limit σ_p. The part of the slab that is at the yield stage and that consequently has considerable deformations can therefore be assumed to be subjected to the reaction σ_p. The conditions are then the same (but just upside-down) as for a column under a slab with uniformly distributed load σ_p. Equations (48) and (49) in YLT, page 87, apply, and so we get

$$m + m' = \frac{P}{2\pi}\left(1 - \sqrt[3]{\frac{\sigma_p}{p_c}}\right), \frac{r}{c} = \sqrt[3]{\frac{p_c}{\sigma_p}} \qquad (4.31)$$

The point load P is assumed to be uniformly distributed over a certain area, so the load on the slab here is p_c. For a circular surface of distribution with radius c, we get $p_c = P/\pi c^2$. The yield pattern is a circle with radius r and yield lines to the centre. The circle is a negative yield line, and the radii are positive yield lines. If the slab is only reinforced at the bottom, m′ can be assumed to correspond to the bending strength of the slab without reinforcement, since the negative circular yield line is formed last and, because of the symmetry, simultaneously, so m′ reaches its ultimate value simultaneously over the whole yield circle.

The formula naturally assumes that the slab is so large that the yield circle does not come too near the edge, since other yield patterns would, in that case, be more dangerous.

4.32 Non-plastic soil

In this case, the reaction of the soil is not known, and we seem to be precluded from formulating a calculation when one part of the system is plastic and the other non-plastic. Indeed, to formulate such a calculation we must introduce new assumptions, the legitimacy of which can only be determined on the basis of tests. An important contribution to this difficult problem has been made by Anders Johansson, civil engineer, of Chalmers Tekniska Högskole (Investigations concerning reinforced concrete slabs for runways at airports, Report No. 2, 1946, and Betong, 1947, vol. 3). We shall therefore begin by reproducing this method of calculation and then will discuss its assumptions.

It is assumed that the soil reaction is $p = ky$, where k is the so-called modulus of the reaction of the ground and y is the deflexion.

Under the plastic part of the slab, the surface of deflexion is assumed to be a cone, whereby p is as shown in Fig. 4.321. Within the negative yield line we can consider the slab as a restrained slab with the load $P - p$, where p can be divided into a uniformly distributed load p_r and a conical load distribution $p_o - p_r$. Then, by means of (4.111) and (4.113), we get

$$m + m' = \frac{P}{2\pi}\left(1 - \frac{2}{3}\frac{c}{r}\right) - \frac{1}{6} p_r r^2 - \frac{1}{12}(p_o - p_r) r^2$$

The equations of projection give

$$P - p_r \pi r^2 - \frac{1}{3}(p_o - p_r)\pi r^2 = 2\pi \bar{q} r = 2\pi(m + m')$$

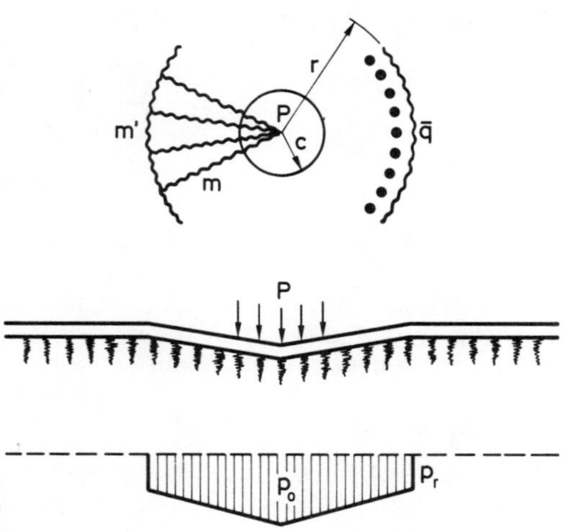

Fig. 4.321

where the transition to the last term corresponds to the equation that is normally valid for $\bar{q}$ (YLT, pages 57 and 82). Eliminating p_r from these two equations by dividing the latter by 8π and subtracting, we get

$$m + m' = \frac{P}{2\pi}\left(1 - \frac{8}{9}\frac{c}{r}\right) - \frac{p_o r^2}{18} \qquad (4.321)$$

We cannot get any further without introducing an assumption about $p_o = k\delta$, where δ is the deflexion under the load. On the basis of model tests at the Institution for Structural Research at Chalmers Tekniska Högskole with 2 to 3 cm thick reinforced concrete slabs on ultra-porous wood fibre-boards (Kramfors) instead of soil, Johansson found that δ can be calculated by means of Hertz's formula for an infinitely large, elastic slab. We then have $p = ky, \Delta\Delta y + p/EI = 0$.

$$p_o = k\delta = \frac{P}{8}\sqrt{\frac{k}{EI}} \qquad (4.322)$$

where EI applies at the yield stage. If the stress-strain curve in bending is OAB, see Fig. 4.322, then EI can be determined by means of the chord OB. The unknown radius r is then determined by means of the maximum principle, which gives $d(m + m') = 0$, i.e.

$$\frac{r}{c} = \sqrt[3]{\frac{4P}{\pi p_o c^2}} = \sqrt[3]{\frac{32}{\pi}}\sqrt[6]{\frac{EI}{kc^4}} \qquad (4.323)$$

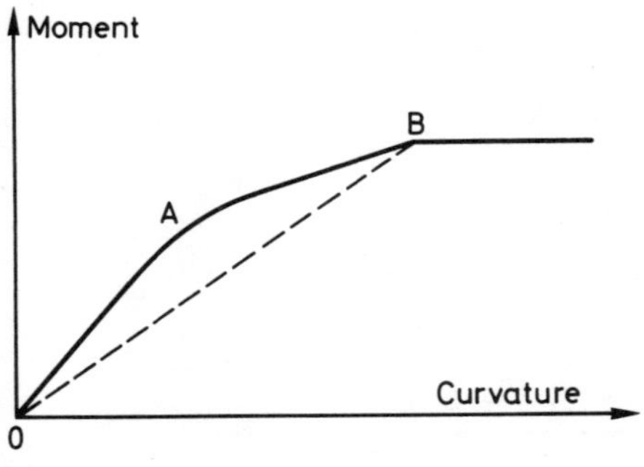

Fig. 4.322

With this value, (4.321) then gives

$$m + m' = \frac{P}{2\pi}\left(1 - \frac{4}{3}\,\frac{c}{r}\right) = \frac{P}{2\pi}\left(1 - \frac{1}{3}\,\sqrt[3]{2\pi}\,\sqrt[6]{\frac{kc^4}{EI}}\right) \tag{4.324}$$

Tests showed that both P and r were in excellent accordance with the values found from these formulae. It is well known that k is dependent upon the size of the loaded area. It is therefore determined as the ratio between the mean pressure and the mean deflexion for the part of the slab that remains in contact with the ground.

To test the formulae in the case of a slab on soil, we can use the slab in Report No. 1 from the above-mentioned investigation. For this slab, $c = 40$ cm, $EI = 16 \times 10^6$ kg cm, $k = 0{\cdot}28$ kg/cm^2, $m = 3300$ kg and $m' = 950$ kg. With these values, we find that $r = 145$ cm and $P = 42{\cdot}2$ t. In the test, r was found to be about 160 cm and $P = 52{\cdot}6$ t. The deflexion δ, calculated by means of (4.322) is $3{\cdot}1$ cm, whilst the measured value is $2{\cdot}2$ cm. These differences show that the formulae are not particularly good for soil. This is hardly surprising when we compare the stress-strain curves for Kramfors boards with those for soil, K and J, respectively, in Fig. 4.323. If, in the foregoing test example, we double the value of EI, we get $r = 145 \times \sqrt[6]{2} = 163$ cm, $P = 40$ t and $\delta = 3{\cdot}1/\sqrt{2} = 2{\cdot}2$ cm. We now have agreement as far as concerns δ and r, but the real ultimate load is 30%, higher than the calculated value; this is, incidentally, known from ordinary slab tests with restrained slabs and can be attributed to the membrane action (see YLT, page 175).

The weak point of the theory is, of course, the determination of p_o or δ by means of (4.322). We might try using (4.118), which gives

$$p_o = k\delta = k\,\frac{mr^2}{2EI} \tag{4.325}$$

Inserting this in (4.321) and using the maximum principle, with $m' = im$, we get

$$m = \frac{P}{2\pi}\;\frac{1 - \frac{8}{9}\,\frac{c}{r}}{1 + i + \frac{kr^4}{36EI}} = \frac{\frac{8}{9}\,\frac{c}{r^2}}{\frac{1}{9}\,\frac{kr^3}{EI}} = \frac{8cEI}{kr^5} \tag{4.326}$$

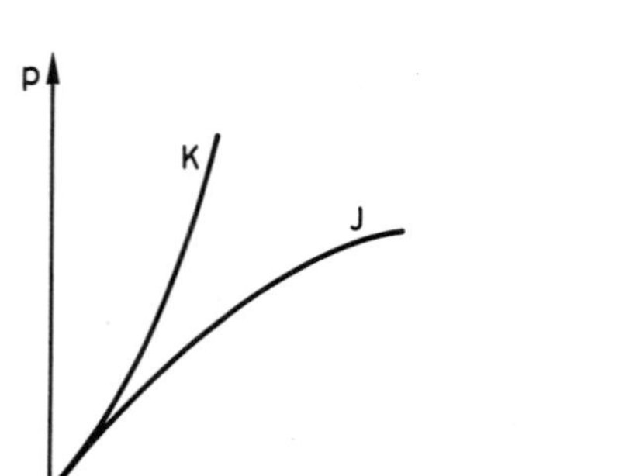

Fig. 4.323

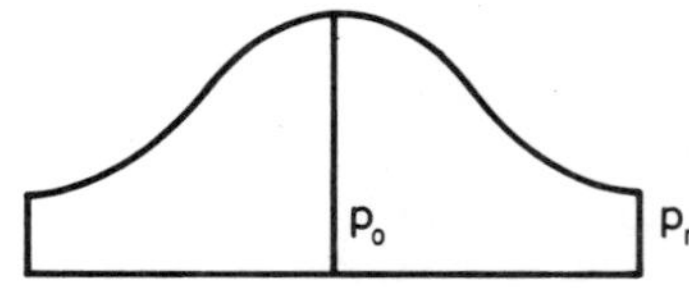

Fig. 4.324

This gives

$$\frac{r^4}{c^4}\left(\frac{r}{c} - \frac{10}{9}\right) = \frac{8EI}{kc^4}\,(1 + i) \tag{4.327}$$

and

$$m + m' = \frac{P}{2\pi}\left(1 - \frac{10}{9}\,\frac{c}{r}\right) \tag{4.328}$$

This gives slightly lower values of r and P than the foregoing and must therefore be regarded as less suitable than these. A distribution of p as shown in Fig. 4.234 would give better agreement, but tests are necessary to establish the value more exactly. By considering the formulae developed we find that we can assume that

$$m + m' = \frac{P}{2\pi}\left(1 - \alpha\,\frac{c}{r}\right),\; \frac{r}{c} = \beta\,\sqrt[6]{\frac{EI}{kc^4}} \tag{4.329}$$

where α and β must be determined by means of tests.

A study of the tests mentioned shows that, for Kramfors boards, β lies close to the theoretical value $\sqrt[3]{32/\pi} = 2{\cdot}17$, whilst α varies from the theoretical value 4/3 to 3. In the case of the slab on soil, we find that $\beta = 2{\cdot}4$ and $\alpha = 2$. It thus appears that β remains almost constant and near the theoretical value, whilst α may be considerably higher, presumably on account of the membrane action. (See also Report No. 3, 1948, from Chalmers.)

4.4 Slab supported on columns

In the following, symmetrical loading is assumed.

The slab is assumed to be free at both edges and loaded with a uniformly distributed load p and a line load $\bar{p}$ at a distance r from the centre (see Fig. 4.41). The columns are assumed to be located symmetrically, so that they lie at the corners of a regular polygon. The axes of rotation must thus also form a regular polygon, and the yield lines will then be the diameters of this. With n columns, we get the column reaction $S = P/n$, where P is the total load. The moment equation about an axis through the centre and parallel with the axis of rotation gives, for one part of the slab (Fig. 4.41):

$$2m(a - b)\sin\alpha = sS - \frac{2}{3}\,p(a^3 - b^3)\sin\alpha + 2\bar{p}r^2\sin\alpha$$

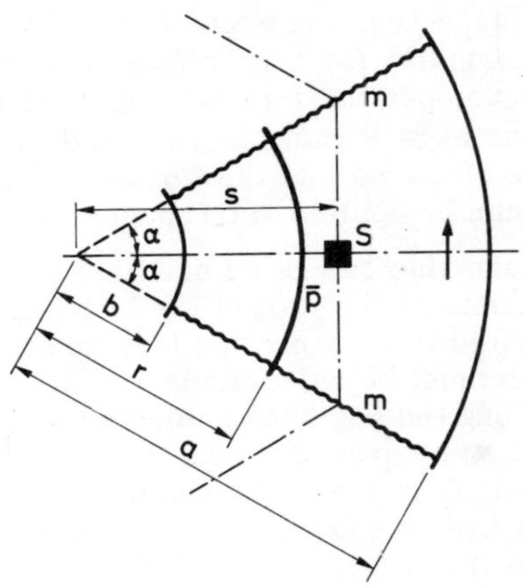

Fig. 4. 41

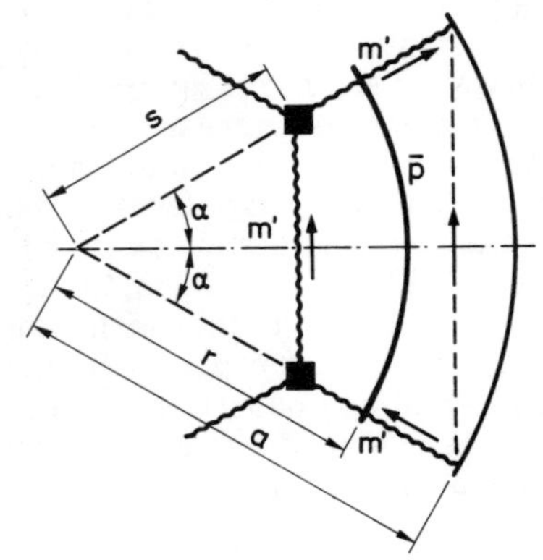

Fig. 4. 42

Here, $\alpha = \pi/n$ and $nS = \pi\, p(a^2 - b^2) + 2\,\pi\, \bar{p}r$, so we get

$$m = \left(p\,\frac{a+b}{2} + \bar{p}\,\frac{r}{a-b}\right) s\,\frac{\pi/n}{\sin(\pi/n)} - \frac{1}{3}\,p(a^2 + ab + b^2) - \bar{p}\,\frac{r^2}{a-b} \qquad (4.41)$$

The moment will be zero when

$$s = \frac{\sin(\pi/n)}{\pi/n} \times \frac{\frac{1}{3}\,p(a^2 + ab + b^2) + \bar{p}\,\frac{r^2}{a-b}}{\frac{1}{2}\,p(a+b) + \bar{p}\,\frac{r}{a-b}} \qquad (4.42)$$

If the moment is found to be negative, the following formulae will give higher values, so (4. 41) only applies for positive moments.

The negative moment is determined by means of the yield pattern in Fig. 4. 42, which is formed by yield lines from column to column and radii from the columns to the edge. The moment equation about the line between the columns for the slab part shown gives

$$2m'\,a \sin\alpha = p(a^2\alpha - s^2 \sin\alpha\cos\alpha)\,t + 2\bar{p}r\alpha\left(r\,\frac{\sin\alpha}{\alpha} - s\cos\alpha\right)$$

The distance t to the centre of gravity is found by

$$(a^2\alpha - s^2\sin\alpha\cos\alpha)(t + s\cos\alpha) = \frac{2}{3}\,a^3\sin\alpha - \frac{2}{3}\,s^3\sin\alpha\cos^2\alpha$$

whence

$$m' = \frac{1}{6}\,pa^2\left(\frac{s^3}{a^3}\cos^2\frac{\pi}{n} + 2 - 3\,\frac{s}{a}\,\frac{\pi}{n}\cot\frac{\pi}{n}\right) + \bar{p}r\left(1 - \frac{s}{a}\,\frac{\pi}{n}\cot\frac{\pi}{n}\right) \qquad (4.43)$$

This formula only applies when $r \geqslant s$. When $s\cos(\pi/n) < r < s$, the contribution from $\bar{p}$ is found to be $2\bar{p}r\beta\,[r\,(\sin\beta/\beta) - r\cos\beta]$, where $r\cos\beta = s\cos\alpha$ (Fig. 4. 43). The contribution from $\bar{p}$ must then be replaced in formula (4. 43) by

$$\bar{p}\,\frac{r^2}{a}\,\frac{\sin\beta - \beta\cos\beta}{\sin\alpha} \qquad (4.44)$$

where $\alpha = \pi/n$ and $\cos\beta = \frac{s}{r}\cos\alpha$.

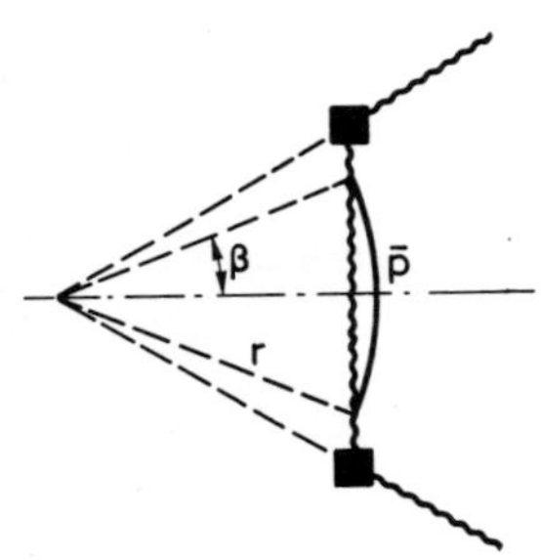

Fig. 4. 43

If there is a hole in the slab, the yield pattern consisting only of the radii through the columns may produce a higher value of m′. It will be seen that the slab parts are then as in Fig. 4. 41, but the column reaction gets the moment arm $s\cos\alpha$. We can therefore immediately use (4. 41), which gives

$$m' = \frac{1}{3}\,p(a^2 + ab + b^2) + \bar{p}\,\frac{r^2}{a-b} - \left(p\,\frac{a+b}{2} + \bar{p}\,\frac{r}{a-b}\right) s\,\frac{\pi}{n}\cot\frac{\pi}{n} \qquad (4.45)$$

when s is replaced by $s\cos\alpha$.

The deflexion at the middle is $EIu = \frac{1}{2}\,ms^2$. The deflexion at the edge is a maximum at the points furthest from the columns. However, in cases in which the deflexion at the edge is critical, we can utilize a strip through a column as in Fig. 4. 212. This gives $EIu = \frac{1}{2}\,m'\,(a^2 - s^2)$.

The capitals have been neglected in the foregoing. This is correct for the positive moment found by means of (4. 41), but not for the negative moment. When the columns have capitals, corresponding yield patterns can be used, where the yield lines now touch the capitals. The formulae developed can then still be used, but angles and distances must correspond to the new yield pattern. Thus, for example in the case of circular capitals with radius c, we find that s in (4. 43) must be replaced by s + c sec (π/n), while the remaining quantities remain unaltered. In (4. 45), s is replaced by $\sqrt{s^2 - c^2}$, and the angle π/n is reduced by arcsin (c/s).

The conditions at the columns must be investigated in the usual way. If they stand at some distance from the edge, the investigation must be made on the basis of the formulae for inner columns. If, on the other hand, they stand near the edge, it must be on the basis of the formulae for edge columns. These have only been developed for straight edges, so when we have a circular edge, it must be replaced by an edge of a suitable shape so that the formulae for corner columns can be applied (YLT, pages 123-124).

However, these formulae cannot be used for the bottom slabs of tanks as the wall of the tank supports the edge, so the yield lines cannot be formed here. Such bottom slabs must be calculated as a circular slab supported along the edge and subjected to a column reaction acting upwards and a downward directed fluid load. Because of the symmetry, the column reactions are already known, since they are 1/n of the total load (fluid load + load from the walls). For the calculation of this type of slab, reference is therefore made to the following chapter.

5 Flat-slab structures

Just as, in the case of continuous beams and one-way slabs, we use the theory of plasticity in order to avoid heavy haunches at the intermediate supports, so, in the case of flat-slab structures, we use the yield-line theory in order to avoid heavy capitals and a local increase in the thickness of the slab over the columns. In the following, therefore, we will only deal with flat-slab structures with modest capitals, no local strengthening, and the simplest possible reinforcement, viz. bottom reinforcement throughout the slab and top reinforcement at the columns only (the latter reinforcement may, as a special case, be a simple ring reinforcement).

The division of the panels into strips used earlier for the design of flat slabs must be considered rather questionable since this entails a very irregular distribution of the reinforcement, some zones getting far too much and others too little. It is particularly characteristic of a number of these methods that the top reinforcement is generally not carried far enough into the panels.

5.0 Conditions at the columns

The conditions at the columns are of considerable importance in flat-slab structures. We will therefore begin by using formulae (48) to (53) in YLT, pages 87 to 88. In the case of square panels with side length l, the column reaction is $S = pl^2$, and the same formula can be used in the case of rectangular panels with the side lengths l_1 and l_2 if we put $l^2 = l_1 l_2$. If we assume the capital to be circular, with radius c, we get $p_c = S/(\pi c^2)$, and the moments at the column, according to formula (49), are

$$m_s + m_s' = \frac{pl^2}{2\pi}\left(1 - \sqrt[3]{\pi \frac{c^2}{l^2}}\right) \qquad (5.01)$$

which can, to a good approximation, be written in the simpler form,

$$m_s + m_s' = \frac{pl^2}{2\pi(1 + 5\frac{c}{l})} \quad \text{when } \frac{c}{l} \leqslant \frac{1}{5} \qquad (5.02)$$

Outside the zone at the column, reinforcement is only provided at the bottom, corresponding to $m = m_s$. We therefore introduce in (53): $m_a = m_s$, $m_a' = m_s'$, $m' = 0$ and $m = m_s$, since there is no local thickening of the slab. With these values, we get

$$R_1 = r,\ R_2{}^2 = \frac{2}{p}\left(\frac{pl^2}{2\pi} - m_s\right),\ a = \frac{p}{3}\,\frac{R_2{}^3 - r^3}{m_s'}$$

where a is the radius to the circle limiting the top reinforcement corresponding to m_s'.

If we insert $m_s = \alpha m_s'$, these formulae can be re-written as follows by means of (5.01):

$$\frac{R}{l} = \sqrt{\frac{1 + \alpha\sqrt[3]{\pi\frac{c^2}{l^2}}}{(1+\alpha)\pi}} \qquad (5.03)$$

$$\frac{a}{l} = \frac{2}{3}\,\frac{\left(1 + \alpha\sqrt[3]{\pi\frac{c^2}{l^2}}\right)\frac{R}{l} - (1+\alpha)\frac{c}{l}}{1 - \sqrt[3]{\pi\frac{c^2}{l^2}}} \qquad (5.04)$$

When a is thereby calculated for various values of α and c/l, we find that $(a - c)/l$ only varies between $1/3$ and $1/4$ as long as $\alpha > 1/3$ and $c/l \leqslant 1/5$. It will therefore be on the safe side to carry the top reinforcement a distance outside the capital corresponding to one-third of the side length l, i.e.

$$a = c + \frac{1}{3}\,l, \text{ for } c \leqslant \frac{1}{5}\,l,\ m_s' < 3m_s \qquad (5.05)$$

The radius R of the outer yield line is generally found to be slightly less than $1/2$ l.

A corresponding investigation shows that the bottom reinforcement must, to all intents and purposes, be taken right out to the capital.

In the derivation of equations (48) to (53) of YLT, the slab is assumed to be isotropically reinforced. However, as we have only assumed radial, negative yield lines, the reinforcement for m_s' can just as well be designed as rings (or a spiral) spaced at constant intervals, whereby 50% of the top reinforcement can be saved. In the following, we shall need the moment resultant for a chord section, and we shall therefore derive a formula for this.

In Fig. 5.0, the chord AB is determined by the angle

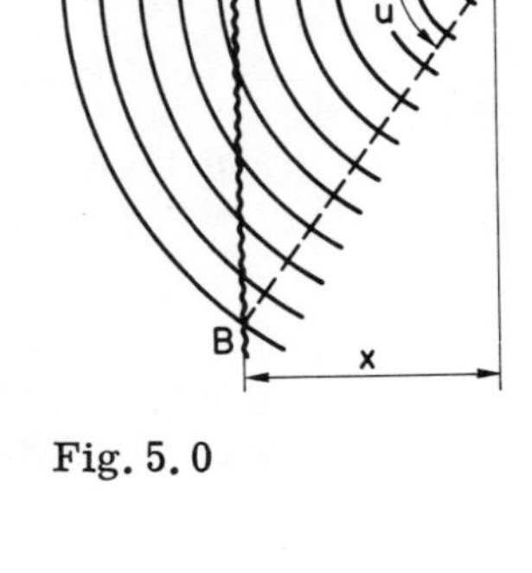

Fig. 5.0

2u. The chord intersects the ring reinforcement at different angles. At C we get the contribution $m'_s \sin^2 v \times dy$ (YLT, pages 52-53), where $dy = (x/\cos^2 v)\, dv$, as $y = x \tan v$. For the whole chord, we therefore get the moment,

$$M_R = 2xm'_s \int_0^u \tan^2 v\, dv = 2\,(\tan u - u)xm'_s$$

$$= 2(\sin u - u \cos u)\, am'_s \qquad (5.06)$$

It should be noted in this connexion that bar reinforcement corresponding to m'_a at the limit, i.e. $rm'_r = am'_a$, gives the contribution $m'_r \cos^2 v\, dy = m'_r x\, dv$ at C. As $r = x/\cos v$, we get the contribution of the bar reinforcement for the whole chord,

$$M_S = 2 \int_0^{\alpha} \frac{am'_a}{x} \cos v\; x\, dv = 2am'_a \sin u$$

$$= \overline{AB} \times m'_a \qquad (5.07)$$

i.e. the same contribution as if the slab were isotropically reinforced, corresponding to m'_a. Reducing m'_a to $\frac{1}{2}m'_a$ at the circle with radius $\frac{1}{2}a$, we get

$$M_S = \left(\sqrt{a^2 - x^2} - \frac{1}{2}\sqrt{\frac{a^2}{4} - x^2}\right) m'_a$$

It is normally the moments that determine the slab thickness and not the shear stress. The thinnest slab is obtained when $m_s = m'_s$. Then, the slab with the stresses 75 and 1300 kg/cm² has a thickness of

$$h \sim 0,3\sqrt{\frac{pl^2}{4\pi\left(1 + 5\frac{c}{l}\right)}}$$

from which we get the shear stress,

$$\tau \sim \frac{pl^2}{2\pi ch} \sim 2\left(5 + \frac{1}{c}\right)\sqrt{p} \qquad (5.08)$$

where p is in kg/cm². With p in kg/m², we get

$$\tau \sim \frac{1}{10}\left(1 + \frac{1}{5c}\right)\sqrt{p} \qquad (5.09)$$

which shows that the shear stress only becomes important at high values of p.

5.1 Inner panels

We will assume that there are so many identical panels in both directions (at least three) that the calculations can be carried out as though there were an infinite number of identical panels in both directions.

5.11 Square panels

We will assume ring reinforcement for m'_s. The yield pattern shown in Fig. 5.111 gives the following

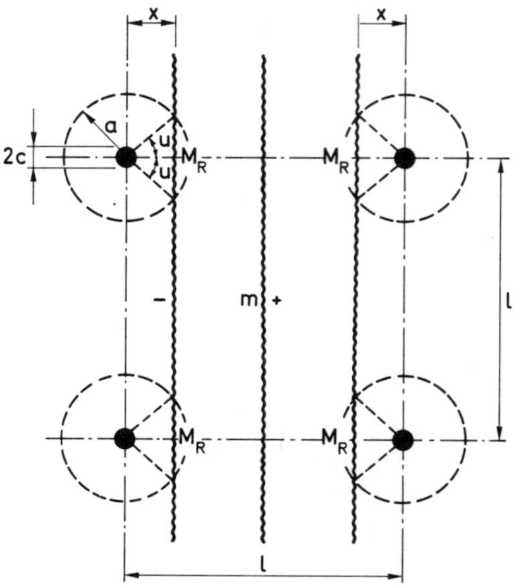

Fig. 5.111

moment equation about the negative yield line for a panel:

$$ml + M_R = \frac{1}{2}\, p\left(\frac{l}{2} - x\right)^2 l$$

or, by means of (5.06):

$$m + 2\,\frac{a}{l}\,(\sin u - u \cos u)m'_s = \frac{1}{8}\, p(1 - 2a \cos u)^2 \qquad (5.111)$$

Differentiating with respect to u, the maximum principle now gives

$$um'_s = \frac{1}{4}\, pl(1 - 2\,a \cos u) \qquad (5.112)$$

From these two equations, we obtain the following for the determination of u:

$$\frac{u}{2} - \frac{m}{m'_s} = \frac{a}{l}\,(2 \sin u - u \cos u) \qquad (5.113)$$

The right side is always positive as $u < 90°$, whereas the left side is only positive for small values of m/m'_s, so this solution is of no practical interest. Maximum values are therefore obtained with the yield lines as close to the columns as possible, i.e. for $x = c$. Then u is determined by means of $c = a \cos u$, and we find

$$m = \frac{1}{8}\, p(1 - 2c)^2 - 2m'_s\left(\frac{a}{l} \sin u - u\,\frac{c}{l}\right) \qquad (5.114)$$

where

$$a = c + \frac{1}{3}\, l, \quad \cos u = \frac{c}{a}$$

If we determine the contribution due to m'_s for values of c/l varying from zero to 1/5, we find that, with a maximum error of 4%, it can be replaced by $\frac{2}{3}\,[1-(c/l)]\,m'_s$, so

$$m = \frac{1}{8}\,p(l-2c)^2 - \frac{2}{3}\left(1-\frac{c}{l}\right)m'_s \text{ for } \frac{c}{l} \leq \frac{1}{5} \quad (5.115)$$

If either m'_s or the ratio $\alpha = m_s/m'_s$ has been fixed in advance, both m'_s and m_s are determined by (5.02). If the value of m determined by means of (5.115) is lower than m_s, the yield pattern in (5.111) cannot be formed, and the calculation must be made on the basis of (5.02). If, on the other hand, we put $m = m_s$ in (5.115), both m_s and m'_s can be determined by means of (5.02) and (5.115). If we do this, we find that, when c/l varies between 0·05 and 0·15, m'_s lies between $pl^2/14$ and $pl^2/15$, whilst $\alpha = m_s/m'_s$ varies between 0·8 and 0·3.

However, for practical reasons, it is not always possible to use these values, and in the cases in which m exceeds m_s, bar reinforcement at the top can be arranged over the column. According to (5.07), this makes the contribution $m'_a\, 2\sqrt{a^2-c^2}$ to the equilibrium equation, and we then get

$$2m'_a\sqrt{a^2-c^2} = (m-m_s)l$$

or, with $a = \frac{1}{3}l + c$,

$$m'_a = \frac{3}{2}\,\frac{m-m_s}{\sqrt{1+6\frac{c}{l}}} \quad (5.116)$$

It will as a rule be necessary to let m'_a decrease to $m'_a/2$ at $\frac{1}{2}a$. We then get

$$2m'_a\left(\sqrt{a^2-c^2} - \frac{1}{2}\sqrt{\frac{a^2}{4}-c^2}\right) = (m-m_s)l \quad (5.116')$$

When $m < m_s$, we *must* reckon with m_s because formula (5.02) assumes this.

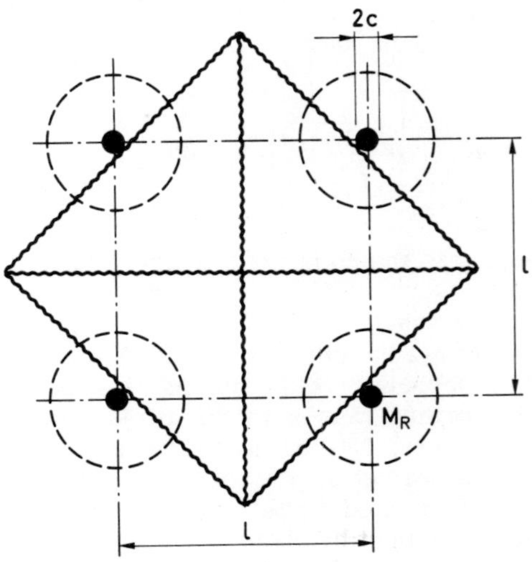

Fig. 5.112

The yield pattern in Fig. 5.112 gives

$$m(l\sqrt{2}-2c) + M_R = \frac{p}{6}\left(\frac{l\sqrt{2}-2c}{2}\right)^2 (l\sqrt{2}-2c)$$

or

$$m + \frac{M_R}{l\sqrt{2}-2c} = \frac{p(l\sqrt{2}-2c)^2}{24}$$

which always gives a lower value of m than (5.115).

In the case of steel mesh reinforcement and similar, we get isotropic reinforcement over the columns. It is then natural to arrange the mesh in square panels with side length 2a over the columns, where a can now be slightly smaller, viz.

$$a = 0{\cdot}3\,l + c \quad (5.05')$$

Corresponding to Fig. 5.111, we then get for x = c,

$$ml + 2am'_s = \frac{1}{8}p\,(l-2c)^2 l$$

or

$$m = \frac{1}{8}\,p(l-2c)^2 - 2\,\frac{a}{l}\,m'_s \quad (5.117)$$

Further, for x = a, we get

$$m = \frac{1}{8}\,p(l-2a)^2 \quad (5.118)$$

The yield pattern corresponding to Fig. 5.112 is of no interest here either.

If we insert $m = m_s$ in (5.117) or (5.118), m_s and m'_s can be determined in connexion with (5.02). It is then found that only the lowest values of c/l give practicable solutions, so (5.02) alone will be decisive in most cases.

This always applies when $m_s = m'_s$, which gives the thinnest slab, in the case of both polar reinforcement and isotropic top reinforcement at the columns. In this case, the slab is calculated solely by means of (5.02). If, for example, the requirements to stiffness call for a thicker slab, a saving in steel can be attained by selecting as high a value of m'_s as possible, i.e. by using both (5.02) and (5.115) or the corresponding formulae for determination of m'_s and m_s.

5.12 *Rectangular panels*

As mentioned, formulae (5.02) and (5.05), with $l^2 = l_1 l_2$, apply for these slabs. When $l_1 > l_2$, we get, corresponding to (5.115),

$$m = \frac{1}{8}\,p(l_1-2c)^2 - \frac{2}{3}\left(\sqrt{\frac{l_1}{l_2}} - \frac{c}{l_2}\right)m'_s \quad (5.121)$$

since the yield pattern corresponding to Fig. 5.111 now gives the equation,

$$ml_2 + M_R = \frac{1}{8}\,p(l_1-2c)^2 l_2$$

where M_R can still be assumed to be

$$M_R = 2m'_s\,(a \sin u - uc) \sim \tfrac{2}{3}\,m'_s\,(1-c)$$

We proceed as in the case of square panels. If we put $m = m_s$, then m_s and m'_s are determined by (5.02) and (5.121). If, on the other hand, we select a value of m'_s or the ratio m_s/m'_s, then m is found from (5.121), and if the value arrived at is smaller than m_s, (5.02) only is decisive. If, on the other hand, it is greater than m_s, bar reinforcement must be provided at the top, and this is determined by

$$m'_a = \frac{m - m_s}{2\sqrt{a^2 - c^2}}\, l_2 = \frac{3}{2}\frac{m - m_s}{\sqrt{\frac{l_1}{l_2} + 6\frac{c}{l_2}\sqrt{\frac{l_1}{l_2}}}} \qquad (5.122)$$

corresponding to (5.116). When m'_a is reduced to $\frac{1}{2}m'_a$ at $\frac{1}{2}a$, we get

$$m'_a = \frac{(m - m_s)\, l_2}{2\sqrt{a^2 - c^2} - \sqrt{\frac{a^2}{4} - c^2}} \qquad (5.122')$$

With isotropic reinforcement at the column within the square with the side 2a, where $a = 0{\cdot}3\, l + c$, we get, corresponding to (5.117) and (5.118),

$$m = \frac{1}{8} p(l_1 - 2c)^2 - 2\frac{a}{l_2} m'_s \qquad (5.123)$$

and

$$m = \frac{1}{8} p(l_1 - 2a)^2, \text{ where } a = 0{\cdot}3\sqrt{l_1 l_2} + c \qquad (5.124)$$

The deflexion is determined by means of a diagonal strip with the span d — 2c:

$$EIu = \frac{1}{8} m_s (\sqrt{l_1^2 + l_2^2} - 2c)^2 \qquad (5.125)$$

5.2 Outer panels

If the outer panel is supported by walls along one or two sides, we can generally only regard the slab as simply supported along such walls. In frame structures, the outer columns can be connected by beams, whereby the slab can be considered as simply supported. If the outer columns are designed so that they can resist the moment M_R from the column capital, the outer panels can be designed in the same way as the inner panels. In the following, therefore, we will only deal with outer panels with simple support.

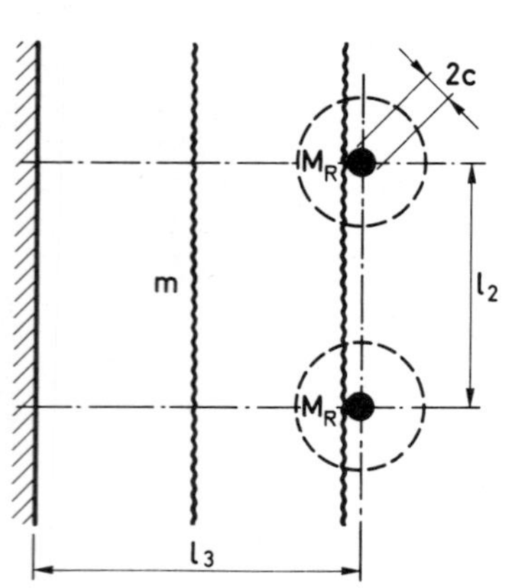

Fig. 5.21

5.21 *Edge panel*

The yield pattern in Fig. 5.21 shows that the panel can be treated as a panel with the span $l_3 - c$ and the restraining moment M_R/l_2 at one side, i.e.

$$m_R \sim \frac{1}{8} p(l_3 - c)^2 - \frac{1}{2}\frac{M_R}{l_2}$$

or

$$m_R = \frac{1}{8} p(l_3 - c)^2 - \frac{1}{3} m'_s\left(1 - \frac{c}{l}\right)\frac{l}{l_2} \qquad (5.211)$$

if polar reinforcement is adopted over the columns. If there is also bar reinforcement, we get the further reduction $\frac{1}{2}(m - m_s)$. Correspondingly, with isotropic reinforcement, we get

$$m_R = \frac{1}{8} p(l_3 - c)^2 - \frac{a}{l_2} m'_s \qquad (5.212)$$

and

$$m_R = \frac{1}{8} p(l_3 - a)^2 \qquad (5.213)$$

l_3 can be determined by means of these formulae in such a way that the edge panels get the same value of m as the inner panels. Corresponding formulae are obtained for the edge panels corresponding to l_1 and l_4. For bar reinforcement, we now get the reduction $\frac{1}{2}(m - m_s)(l_2/l_1)$. The formulae show that the edge panels must in all cases have spans that are somewhat more than the spans of the inner panels reduced by c, i.e. $l_3 < l_1 - c$, $l_4 < l_2 - c$.

With $m'_s = m_s$, $l_1 = l_2$ and polar reinforcement, (5.211), in connexion with (5.02), gives

$$\frac{l_3}{l} = 0.46\sqrt{\frac{4 - \frac{c}{l}}{1 + 5\frac{c}{l}}} + \frac{c}{l} \sim 0{\cdot}85 \qquad (5.214)$$

corresponding to the conditions in one-way slabs.

5.22 *Corner panels*

Corner panels get the above-mentioned spans l_3 and l_4. However, only yield patterns as shown in Fig. 5.22 can be formed, that is, as for a triangular slab with corner levers. Neglecting the restraint at the column capital, we get $m = \frac{1}{6} pr^2$, where r is the radius of the inscribed circle. The corner levers increase this moment by about 25% (see the Table on page 67), so $m = pr^2/4{\cdot}8$. The largest inscribed circle is the one that touches the capital and the sides enclosing the corner. The radius of this circle is determined by $(r + c)^2 = (l_3 - r)^2 + (l_2 - r)^2$, i.e.

$$r = l_3 + l_4 + c - \sqrt{2 l_3 l_4 + 2c(l_3 + l_4 + c)} \qquad (5.221)$$

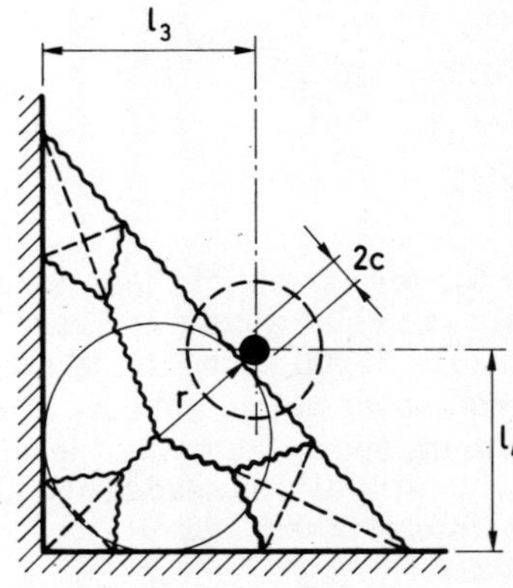

Fig. 5.22

By making use of the work equation for yield patterns without corner levers, it will be seen that the contribution from the moment M_R from the capital is M_R/s, where s is the periphery of the triangle. Because of the corner levers, s must be replaced by a somewhat lower value, and if, for the sake of simplicity, we take $2(l_3 + l_4)$, we get the following estimated value, which is on the safe side:

$$m_H \sim \frac{pr^2}{4 \cdot 8} - \frac{1}{3} m_s' \frac{a}{l_3 + l_4}\left(1 - \frac{c}{l}\right) \qquad (5.222)$$

This value will always be considerably lower than those found earlier, so the corner panels do not affect the design. However, we cannot reinforce for this smaller value of m as the conditions at the column expressed by (5.02) require m_s over most of the panel.

Numerical example

A 15·6 × 33·6 m floor with a load of 1200 kg/m² is to be supported by a flat-slab structure. It is decided to have two rows of columns in the longitudinal direction. A distance between the rows of 6·0 m is selected, so that the outer panels become 4·8 m, which is suitable according to (5.214). With square panels, we get four inner panels and two outer panels of 4·8 m in the longitudinal direction. The corner panels are thus square. The deadweight is estimated at 480 kg/m², so the total load is 1680 kg/m². The column reaction is S = 1680 × 6² = 60480 kg. This results in a column with a side of about 30 cm. A radius of 60 cm for the capital is then suitable. With c/l = 0·1, equations (5.02) and (5.115) give

$$m_s + m_s' = \frac{60480}{3\pi} = 6400 \text{ kg}$$

and

$$m + 0 \cdot 6\, m_s' = \frac{1}{8} \times 1680 \times 4 \cdot 8^2 = 4840 \text{ kg}$$

Selecting $m = m_s$, we find from the above that $m_s' =$ 3900 kg and m_s = 2500 kg, which gives h_n = 16·2 cm and f = 21·7 cm² = 7 × 20 mm dia. bars per m for the stresses 75 and 1300 kg/cm². The slab thus has a thickness of 20 cm, h_n' = 17·0 and f′ = 20·7 cm² = 20 mm dia. bars spaced at 15 cm. Over the column we therefore have to arrange a 20 mm dia. spiral with 15 cm thread or corresponding rings over a length of a = ⅓ × 600 + 60 = 260 cm. Throughout the bottom of the slab we must have 20·7 × (2500/3900) = 13·2 cm² = 20 mm dia. bars spaced at 24 cm.

Using (5.211) for the edge panels, we get $m_R = \frac{1}{8} \times 1680 \times 4 \cdot 2^2 - 0 \cdot 3 \times 3900 = 2530$ kg, i.e. practically the same value as for the inner panels. According to (5.125), with $600\sqrt{2} - 2 \times 60 = 730$ cm as strip length, the deflexion is

$$u = \frac{1}{8}\,\frac{2500 \times 730^2}{2 \times 10^5 \times \frac{1}{12} \times 20^3} = 1 \cdot 3 < \frac{730}{500} = 1 \cdot 5 \text{ cm}$$

5.3 Single row of columns

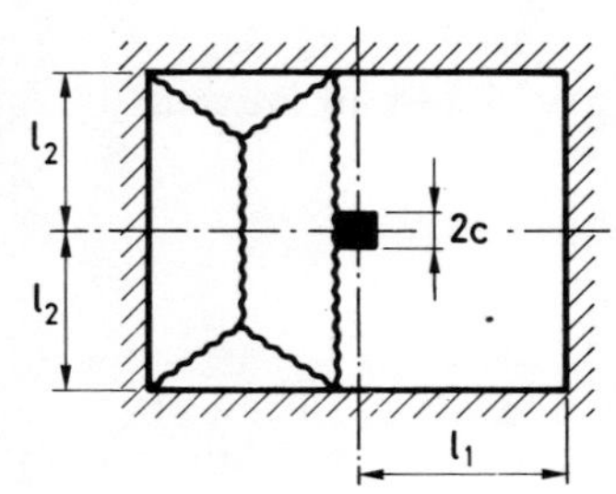

Fig. 5.31

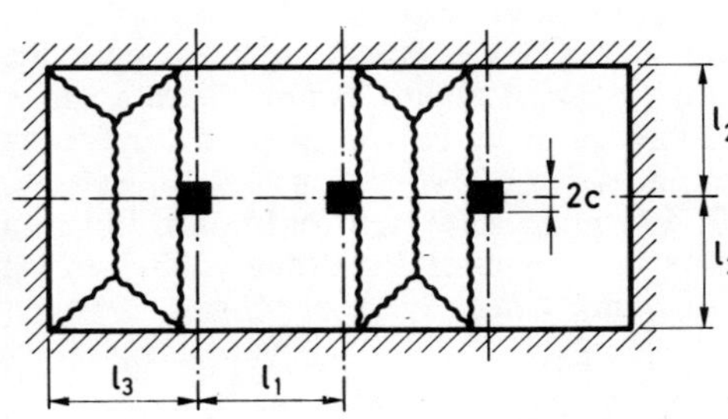

Fig. 5.32

In Figs. 5.31 and 5.32, there are only corner panels and edge panels, and the column reactions are statically indeterminate. In the yield-line theory, the latter condition means that the reactions from the columns are dependent upon the reinforcement, since the reaction from a column must not exceed $2\pi(1 + 5\,c/l) \times (m_s + m_s')$. In other words, the column reaction can be selected within certain limits. It is then natural to select a value about pl_1l_2, corresponding to structures with a number of rows of columns, whereby (5.02) remains valid. Instead of the yield patterns in Fig. 5.111, we get yield patterns corresponding to rectangular slabs, as suggested in Figs. 5.31 and 5.32. Then, by means of the formulae in 1.11, we get

$$m = m_o - m_i = m_o - \frac{M_R}{2(l_3 - c + 2\,l_2)}$$

or

$$m = \frac{p(l_3 - c)\,l_2}{4\left(1 + \frac{l_3 - c}{2l_2} + \frac{2l_2}{l_3 - c}\right)} - \frac{l - c}{3(l_3 - c + 2l_2)}\, m_s' \qquad (5.31)$$

for the outer span in Fig. 5. 32.

In Fig. 5. 31, this is used with $l_3 = l_1$. For the intermediate span in Fig. 5. 32, we get

$$m = \frac{p(l_1 - 2c)l_2}{4\left(1 + \frac{l_1 - 2c}{2l_2} + \frac{2l_2}{l_1 - 2c}\right)} - \frac{2(l - c)}{3(l_1 - 2c - 2l_2)}\, m_s' \qquad (5.32)$$

In the other direction, corresponding formulae can be utilized, but if there is a large number of spans, the formula corresponding to (5.211) can be used:

$$m = \frac{1}{8} p\,(l_2 - c)^2 - \frac{1}{3}\frac{l - c}{l_1}\, m_s' \qquad (5.33)$$

The deflexion can be calculated as follows:

$$EIu = \frac{1}{8}\, m\,(\sqrt{l_1{}^2 + l_2{}^2} - c) \qquad (5.34)$$

Numerical example

The floor dealt with in section 1. 19 is to be supported by a flat-slab structure. With one row of columns, we immediately get $l_2 = 5{\cdot}4$ m. In the other direction, the spans $l_1 = 7{\cdot}0$ m and $l_3 = 5{\cdot}75$ m are adopted. The deadweight is estimated to be 480 kg/m^2, so the total load is 1480 kg/m^2. The value of the column reactions will then be approximately $1480 \times 7 \times 5{\cdot}4 = 56000$ kg, corresponding to a column diameter of about 35 cm. With a view to the span, a radius of 55 cm is adopted for the capital of the column. With $l = \sqrt{7 \times 5{\cdot}4} = 6{\cdot}15$, we get $c/l = 0{\cdot}09$. We must now first calculate the moments in the panels by means of formulae (5. 31) and (5. 32), selecting $m = m_s = m_s'$:

$$m\left(1 + \frac{1}{3}\,\frac{5{\cdot}6}{5{\cdot}2 + 10{\cdot}8}\right) = 1{\cdot}117m$$

$$= \frac{1480}{4}\,\frac{5{\cdot}2 \times 5{\cdot}4}{1 + \frac{5{\cdot}2}{10{\cdot}8} + \frac{10{\cdot}8}{5{\cdot}2}} = 2920 \text{ kg}$$

therefore m = 2610 kg;

$$m\left(1 + \frac{2}{3}\,\frac{5{\cdot}6}{5{\cdot}9 + 10{\cdot}8}\right) = 1{\cdot}224m$$

$$= \frac{1480}{4}\,\frac{5{\cdot}9 \times 5{\cdot}4}{1 + \frac{5{\cdot}9}{10{\cdot}8} + \frac{10{\cdot}8}{5{\cdot}9}} = 3490 \text{ kg}$$

therefore m = 2840 kg;

As there are now only three panels in the length 18·5m, we analyse the portion $(5{\cdot}4 - 0{\cdot}55)\text{m} \times 18{\cdot}5\text{m}$ as a rectangular slab that is restrained at the columns with a moment corresponding to $2M_k$ because there are two columns:

$$m\left(1 + 2 \times \frac{2}{3}\,\frac{5{\cdot}6}{2(18{\cdot}5 + 4{\cdot}85)}\right) = 1{\cdot}16m$$

$$= \frac{1480}{8}\,\frac{4{\cdot}85 \times 18{\cdot}5}{1 + \frac{4{\cdot}85}{18{\cdot}5} + \frac{18{\cdot}5}{4{\cdot}85}} = 3200 \text{ kg}$$

Therefore m = 2760 kg.

If we design the slab for the larger of these values, 2840 kg, it will be able to resist a column reaction of $S = 1{\cdot}45 \times 4\pi \times 2840 = 52000$ kg, which is close to the value calculated above, which moreover corresponds to the maximum span. Inserting the mean span of 6·4 m, we get only 51000 kg instead of 56000 kg, so we can adopt the value of 52000 kg.

The design gives h = 17 cm, and the deflexion is therefore

$$u = \frac{2690 \times 830^2}{8 \times 2 \times 10^5 \times \frac{1}{12} \times 17^3} = 2{\cdot}8 \text{ cm}$$

$$> \frac{830}{500} = 1{\cdot}7 \text{ cm}$$

The slab thickness must be increased to 20 cm.

Over the columns, rings are arranged out to the circle with radius $a = \frac{1}{3} \times 615 + 55 = 260$ cm.

5.4 Bottom of circular tank on columns

In the following we will deal with the case of four columns, because cases with three columns are unlikely to be encountered in practice, although they can be treated in a corresponding manner. Cases with more columns in the same circle differ only slightly from those dealt with in chapter 4.

5.41 Supported on four columns at the middle

The columns are spaced at equal intervals around the circle with radius b. There is an edge load $\bar{p}$ around the circumference, and the bottom can be assumed to be restrained in the wall, corresponding to the restraining moment m_R (positive). The loads on the columns are statically determinate:

$$4S = p\pi a^2 + \bar{p} \times 2\pi a = 4pl^2 \qquad (5.411)$$

whereby the quantity l is determined. The column moments m_s and m_s' are determined by means of (5. 02). Reinforcement is provided over the entire top of the slab.

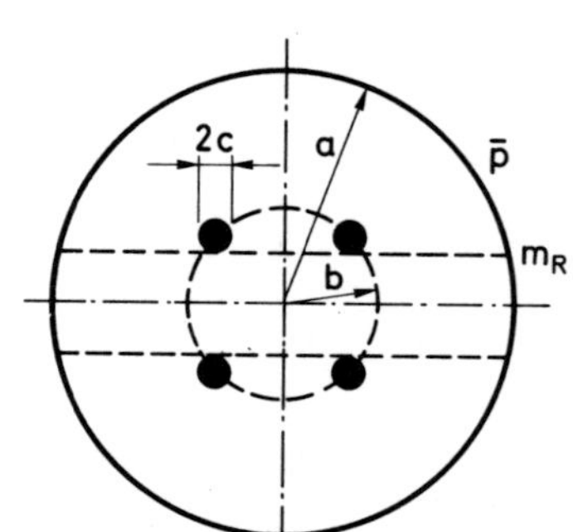

Fig. 5. 41

For the circular ring limited by the radii b + c and a, which is restrained at the inner circle and free at the outer, but here loaded with $\bar{p}$ and the favourable restraining moment m_R, we get, corresponding to (4. 212′), page 78:

$$m_R + m' = \frac{1}{6} p (a - b - c)^2 \left(2 + \frac{b + c}{a}\right) + \bar{p}(a - b - c) \qquad (5.412)$$

The inner circle corresponding to b — c gives, by means of (4. 112), $m + m' = \frac{1}{6}p(b - c)^2$, whilst the rectangle with the sides $b\sqrt{2} - 2c$ and approximately 2a, because of its elongated shape, gives very nearly the greater value

$$m + m' = \frac{1}{8} p (b\sqrt{2} - 2c)^2 \qquad (5.413)$$

As $\bar{p}$ is usually relatively small, $l \sim a$. If we select $c = 0{\cdot}1a$, we then get from (5. 02), $m_s + m'_s = pa^2/12$. Then, using (5. 412), with $m_R = m_s$ and $m' = m'_s$, and neglecting $\bar{p}$, we get $b = 0{\cdot}45a$, whereby (5. 413) gives smaller moments.

If the top reinforcement is only taken out a distance r from the centre, m′ in (5. 412) and (5. 413) must be replaced by m′(r/a). For the determination of r we have partly the fact that the top reinforcement is to be carried a distance of 0·3 l outside the capital (5. 05′), and partly the equation for the circular ring with the radii r and a, m′ now being zero. Analogously with (5. 412), we get

$$m_R = \frac{1}{6} p (a - r)^2 \left(2 + \frac{r}{a}\right) + \bar{p}(a - r) \qquad (5.414)$$

from which r can be determined. The first requirement is generally decisive, i.e.

$$r = b + c + 0{\cdot}3\,l \qquad (5{\cdot}415)$$

We get a slightly higher value of b than above. In the special case above we thus find that $b = 0{\cdot}47a$ and $r = (0{\cdot}47 + 0{\cdot}1 + 0{\cdot}3)a = 0{\cdot}87a$, since $m_s = m'_s = pa^2/24$.

A circular access duct is frequently provided at the centre. This does not alter (5. 412), but (5. 413) no longer applies. Instead, we must consider the inner circular ring with the radius of the duct ρ and outer radius b — c. For this, (4. 221) gives

$$m'(b - c) + m(b - \rho - c) = \frac{1}{6} p(b - \rho - c)^2(b - c + 2\rho) + \bar{p}'\rho(b - c - \rho) \qquad (5.416)$$

where $\bar{p}'$ is the load from the duct. No restraint is assumed in the slender duct. It is assumed that the duct is so small that it has no other effect upon the conditions at the columns than that due to the alteration in the reaction S from the column due to the changed load.

Numerical example

The water tank calculated on page 82 is to be supported on four columns. Estimating the deadweight at 600 kg/m², we get p = 6600 kg/m². The load on the edge is $\bar{p}$ = 4000 kg/m. With a = 4·0 m, we then get from (5. 411):

$$S = \frac{\pi}{4}(6600 \times 4^2 + 2 \times 4 \times 4000) = 108\,000 \text{ kg},$$

$$l = 4{\cdot}05$$

The column reaction gives a column of about 40 × 40 cm, so 60 cm is a suitable value for c. We thereby get $c/l = 0{\cdot}148$, and (5·02) gives

$$m_s = m'_s = \frac{108\,000}{4\pi \times 1{\cdot}74} = 4950 \text{ kg}$$

With the same stresses as on page 82, we get from this $h_n = 20{\cdot}2$ cm, f = 28·8 cm² per m, or 28 mm dia. bars spaced at 21 cm. This again gives h = 25 cm, corresponding to the estimated deadweight. We now have to determine b. If the reinforcement at the top is taken right out to the wall and the lower part of the wall is constructed in the same way as the bottom, whereby $m_R = m_s$, equation (5. 412) yields the following expression for determination of b:

$$2 \times 4950 = \frac{6600}{6} (3{\cdot}4 - b)^2 \left(2{\cdot}15 + \frac{b}{4}\right) + 4000\,(3{\cdot}4 - b)$$

or

$$b - 0{\cdot}93 = (3{\cdot}4 - b)^2\left(0{\cdot}59 + \frac{b}{14{\cdot}6}\right)$$

Therefore b = 2·12 m.

The columns must therefore be spaced at intervals of $2{\cdot}12 \times \sqrt{2} = 3{\cdot}00$ m.

It will be seen from (5. 415) that the reinforcement at the top must be taken out a distance of a least 2·12 + 0·6 + 1·22 = 3·94 m from the centre, i.e. practically all the way to the wall, as assumed.

5.42 *Supported on four columns at the middle and around the circumference*

The column reactions are not statically determinate but must be estimated. A reasonable figure can be obtained by imagining the slab to be supported around the whole circle b and then distributing the reactions thereby determined over the columns. In this way, we get ½ pb from the inner circular part and $\bar{q}_b$ from the outer ring-shaped part, by using (4. 234). We therefore get a total per column of

$$S = \frac{\pi}{2} b\left(\frac{1}{2} pb + \bar{q}_b\right) = \frac{\pi}{8} p(a + b) \frac{m'_a b + m(a + b) + m'_b a}{m'_a + 2m + m'_b} \qquad (5.421)$$

In the special case $m'_a = m'_b$, we get

$$S = \frac{\pi p}{16}(a + b)^2, l = 0{\cdot}443(a + b) \qquad (5.422)$$

This value can also be used when m_b' deviates somewhat from m_a' because S, just as in 5. 3, depends upon the reinforcement.

From (5. 02), we then get

$$m_s + m_s' = \frac{p(a + b)^2}{32\left(1 + 11{\cdot}3\frac{c}{a + b}\right)} \qquad (5.423)$$

For the slab ring determined by means of b + c and a, equation (4. 233) gives

$$m_a' + 2m + m_b' = \frac{p}{15}(a - b - c)^2\left(2{\cdot}75 + \frac{b + c}{a}\right) \qquad (5.424)$$

Further, (5. 413) is valid.

If we assume that $m_a' = m_b' = m_s'$ and that $m = m_s$, equations (5. 423) and (5. 424) yield an equation for the determination of b when c has been chosen. When c varies from 0·05a to 0·2a, we get b = 0·34a to 0·36a. From (5. 423), we can then get values that can be represented by means of

$$m_s + m_s' = \frac{pa^2}{18\left(1 + 7\frac{c}{a}\right)} \quad \text{for } \frac{c}{a} \leqslant 0{\cdot}2 \qquad (5.425)$$

For c = 0, this is on the safe side. The corresponding column reaction, according to (5. 422), is $S = 0{\cdot}36\ pa^2$ when b is assumed to be 0·36a.

It will easily be seen that (5. 413) gives lower moments.

We could also have determined b in such a way that (5. 423) and (5. 413) gave the same value of m + m′, but we would then get b = 0·55a ($0 \leqslant c \leqslant 0{\cdot}1a$) and a considerably higher value of S, and thus also a higher value of m + m′ than above. Formula (5. 424) then gives smaller moments than (5. 423) and (5. 413), so this is a possible solution, although obviously uneconomical. Finally, we could determine b in such a way that (5. 413) and (5. 424) gave the same moments, which means that b = 0·52-0·55a when $0 \leqslant c \leqslant 0{\cdot}2a$, i.e. $m + m' = \frac{1}{8}p(0{\cdot}78a - 2c)^2$, which only gives lower moments than (5. 425) when c > 0·16a, so this formula is not of practical interest either. The column reaction is here determined by means of the moments and is therefore somewhat greater than $0{\cdot}36\ pa^2$.

When there is a hole at the centre, we proceed as in section 5. 41.

These formulae assume isotropic top reinforcement throughout the slab. If we use ring reinforcement over the columns, this will not affect (5. 421) and (5. 422). The limit for ring reinforcement is, as previously, $\frac{1}{3}l + c$. Formula (5. 413) is now altered to

$$m + \frac{2M_R + m_a'(b\sqrt{2} - 2c)}{2a + b\sqrt{2} - 2c} = \frac{p}{8}(b\sqrt{2} - 2c)^2 \qquad (5.426)$$

which is obtained by considering the rectangle with the sides 2a and $b\sqrt{2} - 2c$, the value of m_o being as before, while m_i is determined by $M_R = \frac{2}{3}m_s'(l - c)$ and the restraining moment of the wall, m_a'. With $m_a' = m_s'$, we get, especially,

$$m + \frac{0{\cdot}3a + b - 1{\cdot}7c}{a + 0{\cdot}7b - c}m_s' = \frac{p}{8}(b\sqrt{2} - 2c)^2 \qquad (5.427)$$

The treatment of the panels between the columns and the wall, on the other hand, is much more difficult, because the reinforcement at the top consists partly of rings around each column and partly of bar reinforcement at the wall. However, it is a reasonable assumption that (5. 424) must be approximately applicable provided we replace m_b' by

$$\frac{4M_R}{2\pi(b + c)} \sim 0{\cdot}4\frac{l - c}{b + c}m_s'$$

where the latter expression is slightly on the low side. We also insert M_R with its earlier value, which corresponds to the section along the tangent, although we really ought to use the intersection with the circle with radius b + c, which would give a higher value. Instead of (5. 424), we then get

$$m_a' + 2m + 0{\cdot}4\frac{l - c}{b + c}m_s' = \frac{p}{15}(a - b - c)^2\left(2{\cdot}75 + \frac{b + c}{a}\right) \qquad (5.428)$$

where l is given by (5. 422).

In (5. 423), (5. 426) and (5. 428), we have three equations for determining $m = m_s$, m_s' and b, when the restraining moment m_a' in the wall is known. However, as it is not given that the solution thus obtained is the most advantageous, and as (5. 423) is only based on an estimate, it is best to make a few calculations for some values of b between $\frac{1}{3}a$ and $\frac{1}{2}a$, and select the most favourable of these.

In order to determine how far the reinforcement corresponding to m_a' should be carried into the slab, we consider the circular ring with the radii b + c and r, where r corresponds to the circle where m_a' is zero. For this, an equation corresponding to (5. 428), with $m_a' = 0$ and a = r, applies, i.e.

$$2m + 0{\cdot}4\frac{l - c}{b + c}m_s' = \frac{p}{15}(r - b - c)^2\left(2{\cdot}75 + \frac{b + c}{r}\right) \qquad (5.429)$$

Since the other quantities in this equation are already known, we can find r from it.

Numerical example

A 1000 m^3 water tank has a height of 9 m and a diameter of 12 m, i.e. a = 6 m. The load is p = 9720 kg/m^2. We will try with b = 2·3 m, and we choose c = 0·5 m. With $m_s = m_s' = m_a' = m$, equation (5. 423) gives

$$m = \frac{9720 \times 8{\cdot}3^2}{2 \times 32\left(1 + 11{\cdot}3\frac{0{\cdot}5}{8{\cdot}3}\right)} = 6240 \text{ kg}$$

whilst (5. 427) gives

$$m\left(1 + \frac{1{\cdot}8 + 2{\cdot}3 - 0{\cdot}85}{6 + 1{\cdot}61 - 0{\cdot}5}\right) = \frac{9720}{8}(2{\cdot}3\sqrt{2} - 1{\cdot}0)^2$$

i.e. m = 4230 kg. As l = 0·443 × 8·3 = 3·68, equation (5. 428) gives

$$m\left(3 + 0{\cdot}4\,\frac{3{\cdot}18}{2{\cdot}8}\right) = \frac{9720}{15}\,3{\cdot}2^2\left(2{\cdot}75 + \frac{2{\cdot}8}{6}\right)$$

Therefore m = 6200 kg.

We therefore get practically identical maximum values.

Repeating the calculation with b = 2·5 m, we get m = 6600 kg, 5310 kg and 5530 kg, respectively, where the two lowest values are nearly the same. As the first value is based on the estimated column reaction, we can just as well alter this slightly, and it would even be reasonable to reduce it a little since, in the derivation, we replaced the four columns with a ring-shaped support that naturally gets a slightly heavier load. The slab can then be designed for m = 5530 kg. As l = 376 cm, the limit for the ring reinforcement is $\frac{1}{3}$ × 376 + 50 = 175 cm. By means of (5. 429), we get $(r - 3)^2\,[2{\cdot}75 + (3/r)] = 20{\cdot}8$, r = 5·5, i.e. the reinforcement at the restrained edge must be taken 50 cm into the slab.

6 Choice of reinforcement

It is characteristic of the yield-line theory that, within certain limits, it permits an arbitrary distribution of the reinforcement. In the following, we shall investigate the most economical distribution and the requirements that must be made with a view to the prevention of cracking.

6.1 Isotropic reinforcement—Orthotropic reinforcement

In most of the types of slab investigated, isotropic reinforcement is assumed because it is a relatively simple matter to convert the calculation of orthotropic slabs to that of isotropic slabs by means of the affinity theorem in YLT, page 67 and onwards. If we reinforce a rectangular slab with the sides a and b in such a way that sections parallel with a get the ultimate moment μm, and sections parallel with b, the ultimate moment m, and the same applies to the reinforcement corresponding to the negative moments, the slab will have the same ultimate moment m as the isotropic slab with the sides a and $b/\sqrt{\mu}$, and the same load per unit area.

If the moment for an isotropic slab is m(a, b), where m, as indicated, is a function of a and b, the moment for the orthotropic slab will be $m(a, b/\sqrt{\mu})$. A general investigation of the most economical value of the reinforcement ratio μ is not possible because so many factors are involved besides the required steel and concrete quantities. On the other hand, in cases in which the slab thickness is dependent not upon the moment but, for instance, upon the requirements for stiffness or similar considerations, we can determine the value of μ that gives the smallest amount of steel. When the slab thickness is independent of m, it is also independent of μ, and the quantity of steel is then proportional to

$$J = (1 + \mu) m\left(a, \frac{b}{\sqrt{\mu}}\right) \tag{6.11}$$

Thus, using the exact formula (1.116), which gives simpler calculations than the approximate formulae, we get the following quantity of steel for a rectangular slab that is simply supported along the entire boundary:

$$J = \frac{pa^2}{24}\frac{a^2}{b^2}(1+\mu)\left(\sqrt{3\frac{b^2}{a^2}+\mu} - \sqrt{\mu}\right)^2$$

We thus have to find the minimum of

$$\sqrt{1+\mu}\left(\sqrt{3\frac{b^2}{a^2}+\mu} - \sqrt{\mu}\right)$$

or, which comes to the same thing, the minimum of

$$\frac{\sqrt{1+\mu}}{\sqrt{3\frac{b^2}{a^2}+\mu} + \sqrt{\mu}}$$

which, if we use equation (40) on page 64 of YLT, gives

$$\frac{1}{\mu} = 3\frac{b^2}{a^2} - 2 \tag{6.12}$$

The moment is then

$$m = \frac{1}{8} pa^2\left(1 - \frac{2}{3}\frac{a^2}{b^2}\right) \tag{6.13}$$

The same formulae, with a_r and b_r instead of a and b, apply for restrained slabs.

The result is not of much practical importance, as the saving in steel at the transition to orthotropic reinforcement is slight in relation to the increase in m. Thus, $b = a\sqrt{2}$ and $b = 2a$ give an 8% and a 19% saving respectively but, in both cases, a 48% increase in m. Our assumption that the slab thickness is independent of the moment is therefore only valid in the case of rather thick slabs (the thickness being determined by stiffness requirements or similar); however, such cases seldom arise in practice.

If we now consider the rectangular slab that is simply supported along three sides and free on the fourth, the exact formula (3), page 19, with $a/\sqrt{\mu}$ and b, gives the following (in this case we assume m in sections parallel with a, and μm in sections parallel with b):

$$J = \frac{pab}{8}\frac{1+\mu}{\frac{2}{3}\frac{a}{b} + \sqrt{\frac{4}{9}\frac{a^2}{b^2}+\mu}}$$

which assumes its minimum value at

$$\mu = 1 - \frac{4}{3}\frac{a}{b} \text{ for } \frac{a}{b} \leqslant 0{\cdot}45 \tag{6.14}$$

For this value, we get

$$m = \frac{1}{8} pab \tag{6.15}$$

When the yield pattern is as shown in Fig. 1.212, the slab is calculated as a slab with the sides 2a and b (2a > b), simply supported along the entire boundary. Then (6.12) yields

$$\frac{1}{\mu} = 12\frac{a^2}{b^2} - 2 \text{ for } \frac{a}{b} \geqslant 0{\cdot}55 \tag{6.12'}$$

and the moment is

$$m = \frac{1}{8} pb^2\left(1 - \frac{1}{6}\frac{b^2}{a^2}\right) \tag{6.13'}$$

As the transition from the yield pattern in Fig. 1.211 to that in Fig. 1.212 takes place when the ratio between the sides exceeds 0·71, formulae (6.14) and (6.15) only apply if the ratio between the sides $a/\sqrt{\mu}$ and b in the corresponding isotropic slab does not exceed 0·71, i.e. $\mu \geqslant 2\,a^2/b^2$, which, by means of (6.14), immediately gives the limit indicated in connexion with this formula. In the same way, (6.12′) and (6.13′) only apply for $\mu \leqslant 2\,a^2/b^2$, which, together with (6.12′), gives the indicated limit. In the interval from 0·45 to 0·55, the value of μ that gives the minimum cannot be used for any of the formulae. μ should therefore be determined as the value for which both formulae give the same moment. As the interval is so small, we can, as more detailed calculations will show, use simple interpolation between the values for $a/b = 0{\cdot}45$ and $0{\cdot}55$, thus obtaining

$$\mu = 2\,\frac{a}{b} - 0{\cdot}5, \quad m = \frac{pb^2}{17{\cdot}8} \text{ for } 0{\cdot}45 < \frac{a}{b} < 0{\cdot}55 \qquad (6.16)$$

Using these formulae, we find for b = a, 2a and 3a, a saving in steel of 19%, 7% and 3% respectively, and an increase in moment of 48%, 25% and 25% respectively. It will be seen that savings can only be made in the case of the stiff slab, but the increase in the moment renders this impossible. In the case of flexible slabs, the increase in moment corresponds quite well to the increase required by the stiffness, but the saving in steel is, on the other hand, insignificant.

These and similar investigations show that no major savings in steel can be made. We will therefore conclude this section by drawing attention to a number of cases in which orthotropic reinforcement produces a definite saving without any increase in moment. This applies to the slabs treated in 1.4 and 1.5, with line loads and wheel loads. If the width b of the slabs is so small that the moment $m_b = M/b$, which is found by considering the slab as a beam, is critical, i.e. is greater than m, the slab can be provided with lighter reinforcement in the transverse direction. We then choose the reinforcement ratio in such a way that $m_\mu = m_b$. According to the affinity theorem (YLT, page 68), m_μ is found as the moment in an isotropic slab with the same span and the load $P/\sqrt{\mu}$ or $\bar{p}/\sqrt{\mu}$, i.e. $m_\mu = m/\sqrt{\mu}$, where m has the value determined by the formulae of sections 1.4 and 1.5. From $m_\mu = m/\sqrt{\mu} = m_b$, we then immediately find the ratio of reinforcement,

$$\mu = \left(\frac{m}{m_b}\right)^2 \qquad (6.17)$$

6.2 Polar reinforcement

As we saw in Chapter 4, we can have an arbitrary distribution of the reinforcement between ring reinforcement and bar reinforcement, and it was mentioned that the lowest steel requirement was achieved by using only bar reinforcement, provided the slab thickness was kept constant.

This can be proved by considering the change in the quantity of steel caused by the insertion of a single ring in a slab with only bar reinforcement. The quantity of steel in the bar reinforcement is proportional to the numerical areas of the rm_r curve. The corresponding contribution from the ring is rM_φ, where M_φ is the moment corresponding to the ring, the factor 2π being omitted in both cases.

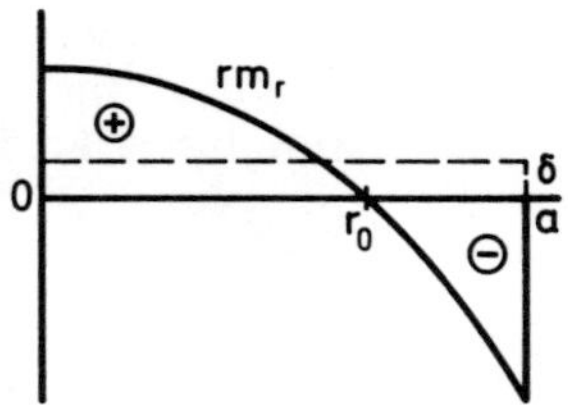

Fig. 6.21

It can be seen directly from equation (4.04) that the insertion of a ring corresponding to M_φ entails a jump of the same magnitude in the rm_r curve.

Fig. 6.21 shows the rm_r curve for a restrained slab. The base line* is located in such a way that the quantity of steel J is as small as possible, i.e. a slight shift δ of the base line should give $\triangle J = 0$. With r_0 as the radius to the point of zero moment, we get a reduction of the positive area and an increase in the negative area, so $\Delta J = -\,\delta r_0 + \delta\,(a - r_0)$ $= \delta\,(a - 2r_0)$, which is zero for $r_0 = \frac{1}{2}\,a$. According to the geometrical conditions on page 74, m_φ must be positive, so a single ring corresponding to $M_\varphi = \delta$ causes the jump in the rm_r curve shown in Fig. 6.22.

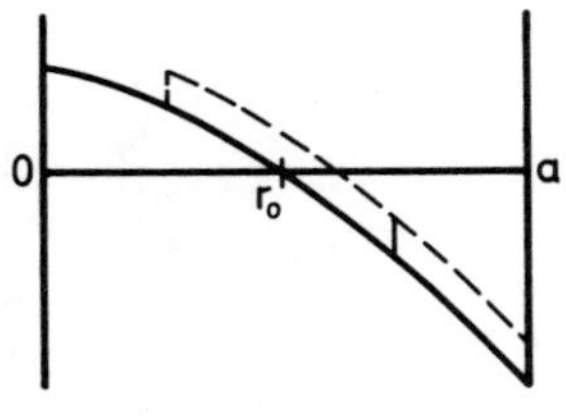

Fig. 6.22

As the ring contributes $r\delta$ to $\triangle J$, and the bar reinforcement contributes $\delta(r_0 - r) - \delta(a - r_0)$ $= \delta(2r_0 - r - a)$ when $r < r_0$, and $-\,\delta(a - r)$ when $r > r_0$, we get $\triangle J = 0$ and $\triangle J = \delta(2r - a)$ respectively. The first result shows that the quantity of steel is not changed by the insertion of ring reinforcement in the inner part of the slab, which behaves as a simply supported slab with positive m_φ and m_r - a result already arrived at (page 75). The other result shows that $\triangle J = 2\,\delta(r - r_0) > 0$ when $r > r_0$, i.e. the quantity of steel is increased by the use of ring reinforcement, and the minimum is consequently obtained using only bar reinforcement.

---

* The values of rm_r are here represented as the ordinates of the curve measured from a certain straight line. This line is referred to in the following as the base line. In the case of Fig. 6.21, the base line is the horizontal line shown in the Figure. (Translator's note.)

As a shift in the base line does not alter J, for r_o was determined in just such a way, we could also have reduced rm_r by δ over the length from the centre to r, whereby we would find that $\Delta J = \delta r - \delta r = 0$.

For a simply supported circular ring, m_φ must be negative at the innermost support and positive at the outermost, as it changes signs at a point in the middle, where the surface of deflexion has a horizontal tangent plane.

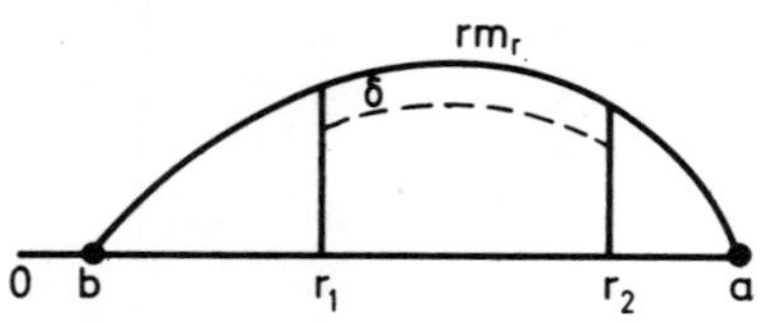

Fig. 6. 23

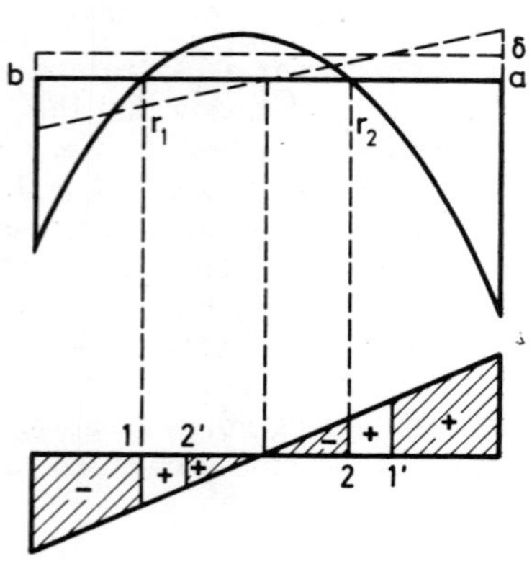

Fig. 6. 24

The above-mentioned theorem derived earlier now no longer applies because m_φ changes signs. If a ring corresponding to $M_\varphi = -\delta$ is placed at r_1 and another ring corresponding to $M_\varphi = +\delta$ at r_2, the edge moments will still be zero because the rm_r curve will be as shown in Fig. 6. 23. The rings increase J by $\delta r_1 + \delta r_2$, whilst the bar reinforcement is reduced corresponding to $\delta(r_2 - r_1)$, so $\Delta J = \delta(r_1 + r_2 - r_2 + r_1) = 2\delta\, r_1 > 0$. The arrangement of rings thus always means an increase in the amount of steel, so the minimum is achieved by using only bar reinforcement.

In the case of the restrained circular ring, the base line is first determined in such a way that the amount of steel is a minimum.

A slight parallel shift δ gives

$$J = \delta(r_1 - b) + \delta(a - r_2) - \delta(r_2 - r_1)$$
$$= \delta\,[a - b - 2(r_2 - r_1)]$$

At the minimum, $\Delta J = 0$, i.e. $r_2 - r_1 = \frac{1}{2}(a - b)$, or the distance between the points of zero moment must be equal to half the span.

By a slight rotation about the mid-point, we find that $\Delta J = 0$ when the points of zero moment are located at equal distances from the middle. At the bottom of Fig. 6. 24 are shown the points 1' and 2', which are symmetrical to the points of zero moment 1 and 2. As the hatched areas cancel out, there will be a positive contribution corresponding to the two equal areas between 1 and 2', and 2 and 1'. Only when these points coincide do we get $\Delta J = 0$.

A comparison with the above result shows that the points of zero moment must lie at the quarter-points, i.e. $r_1 = (a + 3b)/4$, $r_2 = (3a + b)/4$.

As the middle part of the slab may be regarded as a simply supported circular ring, we know from the foregoing that ring reinforcement entails an increase in J. At the inner edge, m_φ is negative, and a ring corresponding to $M_\varphi = -\delta$ gives $\Delta J = \delta r - \delta(r - b) = \delta\, b > 0$. At the outer edge, m_φ is positive, and a ring corresponding to $M_\varphi = \delta$ gives $\Delta J = \delta r - \delta(a - r) = \delta(2r - a) > \delta(2r_2 - a) = \delta(a + b)/2 > 0$. We thus also get an increase in J here if we use ring reinforcement. If the circular ring is simply supported at one of the supports, the base line that gives $\Delta J = 0$ will give the point of zero moment determined by $x = l\sqrt{\frac{1}{2}}$ (Fig. 6. 25), which is found by rotating the base line slightly about the simple support. In a similar manner as above, it can easily be shown that ring reinforcement causes an increase in J.

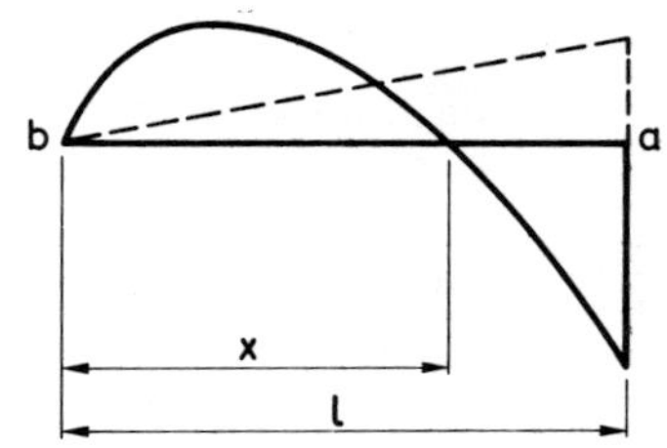

Fig. 6. 25

As a circular slab with symmetrical loading and without ring reinforcement can be calculated in the same way as a beam with $M = rm_r$ and the load pr, as mentioned on page 75, it can be said that the true two-dimensional slab effect is not utilized in such a slab. We have then arrived at the paradoxical result that precisely such a slab gives the lowest amount of steel. As, however, these slabs have very large moments in comparison with slabs with ring reinforcement, this result is principally of theoretical interest, even though certain flexible types of slab may require so great an increase in the thickness that the result can be used for these.

6.3 Degree of restraint

In practice, a degree of restraint of 1/2 is often adopted but, as will be seen from the following, it is advisable to aim at higher values. The degree of restraint is usually of minor importance to the amount of steel, since the saving achieved by means of a high degree of restraint is relatively small. Thus, if we consider a square slab restrained along all sides, the amount of steel is proportional to $ma^2 + m' \times 0{\cdot}64a^2$ when the reinforcement at the restrained edges is taken a distance of 0·2a into the slab. As $m + m' = P/24$, the amount of steel will be proportional to $(1 + 0{\cdot}64i)/(1 + i)$, i.e. proportional to 1, 0·88 and 0·82 for $i = 0, 1/2$ and 1, respectively. There is thus only a slight saving to be made by putting $i = 1$ instead of $i = 1/2$.

Because of the danger of cracking, we should not deviate too much from the values given by the theory of elasticity. If the stress-strain diagram for the slab were straight up to the yield limit, and the factor of safety were 2, we could not use a lower moment of restraint than half the value given by the theory of elasticity if yielding at the working load were to be avoided. However, as the stress-strain diagram is not straight, we can actually deviate somewhat more and, in the absence of more detailed investigations, a value of 1/3 must be considered suitable. Tests with beams have shown that, when half the value given by the theory of elasticity is used, only the usual fine cracks occur, but with a quarter of that value, wide cracks occur at the working load. These tests show that the critical value lies between 1/2 and 1/4 for beams. For slabs, where the maximum moment only occurs at a single point, unlike beams, where the maximum moment occurs over the whole section, conditions are perhaps more favourable.

If we now consider a square slab that is geometrically restrained along all four sides, the theory of elasticity gives the maximum moment of restraint as equal to $P/20$. The restraint has thus weakened the slab, since the maximum moment has increased from $P/24$ in the case of simple support. When we apply the yield-line theory, we must, in accordance with the above rule, not assume a value of m' lower than $P/60$, and as $m + m' = P/24$, we get $m = P/40$, $i = m'/m = 2/3$.

If the slab is only geometrically restrained along one side, the maximum moment of restraint will be $P/12$, i.e. twice the moment with simple support. In the yield-line theory, we must not assume m' to have a value of less than $P/36$, and as $m + \frac{1}{4} m' = P/24$, we get $m = 5P/144$, $i = m'/m = 4/5$.

We see that higher degrees of restraint than 1/2 are required. However, geometric restraint is seldom found in practice. Admittedly, symmetrically loaded continuous slabs have horizontal tangents at the supports, but these are generally beams and thus elastic, whereas the above values assume rigid supports. In the case of elastic supports, the moments of restraint will be smaller, so we can use lower degrees of restraint in the yield-line theory, depending upon the degree of stiffness of the supports.

A rectangular slab with the sides a and 2a, where the sides a are simply supported and the remaining two sides are free and restrained, respectively, has the maximum moment of restraint $0{\cdot}16P$ according to the theory of elasticity. In yield-line theory, we must then have $m' \geqslant 0{\cdot}053P$, which, by means of (1.218) on page 23, gives $i/(9 + 6i) \geqslant 0{\cdot}053$, $i \geqslant 0{\cdot}7$, i.e. once again more than 1/2.

Subject index

Sign convention, signs and symbols

Rotations and moments are represented by vectors along the axis of rotation, in such a way that when one is facing the vector arrow they turn anti-clockwise.

For a yield line, the moment vectors are indicated on the parts of the slab on which the corresponding moments act. The vectors thus form pairs (since they must be drawn slightly apart so that they do not coincide), which with positive yield lines turn anti-clockwise, i.e. in the positive direction of rotation. With negative yield lines, the pair of vectors turn in the negative direction of rotation.

Stresses are indicated in the usual way, i.e. outside the part on which they operate.

Generally, a symbol is included in the following list only if it appears several times in the book, i.e. when the symbol is used in sub-sections other than the numbered sub-section in which it is defined.

Symbol	Meaning	First occurrence on page	Only used on pages
a_r	reduced span of rectangular slab	5	5-19
a_r		20	20-25
b_r		5	5-25
c	radius of capital	87	87-95
E	modulus of elasticity of concrete	15	
g	dead load	2	
H	corner force	6	
i	degree of restraint	2	
I	moment of inertia of a unit length of slab cross-section	15	
J	quantity of steel	77	
k	side length of wheel load	55	55-56
k	ratio between moment with and moment without corner levers	55	55-58
k		59	59-64
k		66	66-67
k		69	69
k		70	70
k′		70	70
k′		70	70
k_x	curvature	15	
l	column distance for square panels	87	87-90
l_1, l_2	column distances for rectangular panels	87	87-90
l_r	reduced span	2	2-4
m	positive ultimate moment	2	
m′	negative ultimate moment	2	
m_o	simple moment	2	
m_i	mean value of restraining moment	6	
m_i		10	
m_i		21	
m_i		58	
m_i		59	
m_i		61	
m_i		91	
m_s	positive ultimate moment in slab near column	87	
m'_s	negative ultimate moment in slab near column	87	

Symbol	Meaning	First occurrence on page	Only used on pages
m_φ	bending moment in radial sections	73	
m_r	bending moment in circular sections	73	
M_R	moment from ring reinforcement	88	88-89
M_R	moment from column capital	89	89-95
M_S	moment from bar reinforcement	89	89-95
p	triangular load	7	
p	uniformly distributed load	5	
$\bar{p}$	uniformly distributed line load	4	
p_c	uniformly distributed column reaction	87	87-95
$\bar{P}$	total line load	10	
P	total distributed load	10	
P	point load	3	
q	uniformly distributed total load	2	
$\bar{q}$	distributed reaction	6	
Q	total distributed reaction on slab side	21	
R	reaction	21	
S	column reaction	41	
u	deflexions	15	
u	angles etc.		
YLT	abbreviation for the book "Yield-Line Theory"	1	
α	ratio between entire line load in the a direction and entire distributed load	8	8-54
α	angles, lengths, etc.	6	
β	ratio between entire line load in the b direction and entire distributed load	8	8-54
β	angles, lengths, etc.	34	
μ	ratio between ultimate moments in orthotropic slab	17	
ω	acute angle between sides of parallelogram-shaped slab	55	55-58
ΔA	correction in work equation for corner lever in A	66	

yield line — column — line load — axis of rotation — restrained edge — point load — free edge — simply supported edge — transverse forces